3rd International Workshop on Rumours and Deception in Social Media (RDSM 2020)

Held online due to COVID-19

Barcelona, Spain
13 December 2020

ISBN: 978-1-7138-2837-2

RDSM 2020

3rd International Workshop on Rumours and Deception in Social Media (RDSM)

Proceedings of the Workshop

December 13, 2020
Barcelona, Spain (Online)

Introduction

Welcome to the 3rd International Workshop on Rumours and Deception in Social Media (RDSM) co-located with the 28th International Conference on Computational Linguistic (COLING 2020), 13th December, 2020.

Abstract

The 3rd edition of the RDSM workshop particularly focuses on online information disorder and its interplay with public opinion formation.

Social media is a valuable resource for mining all kind of information varying from opinions to factual information. However, social media houses issues that are serious threats to the society. Online information disorder and its power on shaping public opinion lead the category of those issues. Among the known aspects are the spread of false rumours, fake news or even social attacks such as hate speech or other forms of harmful social posts. In this workshop the aim is to bring together researchers and practitioners interested in social media mining and analysis to deal with the emerging issues of information disorder and manipulation of public opinion. The focus of the workshop is on themes such as the detection of fake news, verification of rumours and the understanding of their impact on public opinion. Furthermore, we put a great emphasis on the usefulness and trust aspects of automated solutions tackling the aforementioned themes.

Theme and Topics

The workshop brings together researchers and practitioners interested in social media mining and analysis to deal with the emerging issues of veracity assessment, fake news detection and manipulation of public opinion. The workshop put also emphasis on qualitative studies performing user studies on the challenges encountered with the use of social media, such as the veracity of information and fake news detection, as well new data sets. Finally, the workshop welcome studies reporting the usefulness and trust of social media tools tackling the aforementioned problems.

We received 17 submissions, and due to a rigorous review process, we rejected 10. All accepted papers are oral presentations.

Organizers:

Ahmet Aker, University of Duisburg-Essen, Germany and University of Sheffield, UK
Arkaitz Zubiaga, Queen Mary University of London, UK
Kalina Bontcheva, University of Sheffield, UK
Maria Liakata, University of Warwick and Alan Turing Institute, UK
Rob Procter, University of Warwick and Alan Turing Institute, UK
Symeon (Akis) Papadopoulos, CERTH-ITI, GR

Program Committee:

Pepa Atanasova, University of Copenhagen, Denmark
Giannis Bekoulis, Ghent University, Belgium
Costanza Conforti, University of Cambridge, UK
Thierry Declerck, DFKI GmbH, Germany
Leon Derczynski, IT University Copenhagen, Denmark
Samhaa R. El-Beltagy, Newgiza University, Egypt
Genevieve Gorrell, University of Sheffield, UK
Elena Kochkina, University of Warwick, UK
Dominik Kowald, Graz University of Technology, Austria
Chengkai Li, The University of Texas at Arlington, USA
Diana Maynard, University of Sheffield, UK
Preslav Nakov, QCRI, Qatar
Viviana Patti, University of Turin, Italy
Georg Rehm, DFKI GmbH, Germany
Paolo Rosso, Technical University of Valencia, Spain
Carolina Scarton, University of Sheffield, UK
Ravi Shekhar, Queen Mary University of London, UK
Panayiotis Smeros, EPFL, Switzerland
Antonela Tommasel, UNICEN, Argentina
Adam Tsakalidis, Alan Turing Institute, UK
Onur Varol, Northeastern University, USA
Svitlana Volkova, Pacific Northwest National Laboratory, USA

Invited Speaker:

TBA

Table of Contents

Conference Program

Towards Trustworthy Deception Detection: Benchmarking Model Robustness across Domains, Modalities, and Languages

Maria Glenski, Ellyn Ayton, Robin Cosbey, Dustin Arendt, and Svitlana Volkova
Pacific Northwest National Laboratory
Richland, WA, USA
`first.last@pnnl.gov`

Abstract

Evaluating model robustness is critical when developing trustworthy models not only to gain deeper understanding of model behavior, strengths, and weaknesses, but also to develop future models that are generalizable and robust across expected environments a model may encounter in deployment. In this paper we present a framework for measuring model robustness for an important but difficult text classification task – deceptive news detection. We evaluate model robustness to out-of-domain data, modality-specific features, and languages other than English.

Our investigation focuses on three type of models: LSTM models trained on multiple datasets (Cross-Domain), several fusion LSTM models trained with images and text and evaluated with three state-of-the-art embeddings, BERT ELMo, and GloVe (Cross-Modality), and character-level CNN models trained on multiple languages (Cross-Language). Our analyses reveal a significant drop in performance when testing neural models on out-of-domain data and non-English languages that may be mitigated using diverse training data. We find that with additional image content as input, ELMo embeddings yield significantly fewer errors compared to BERT or GLoVe. Most importantly, this work not only carefully analyzes deception model robustness but also provides a framework of these analyses that can be applied to new models or extended datasets in the future.

1 Introduction

Detection of deceptive content online is an extremely important but challenging task and there have been significant efforts to apply machine learning and deep learning to solve this problem (Rubin et al., 2016; Mitra et al., 2017; Wang, 2017; Karadzhov et al., 2017; Volkova et al., 2017; Shu et al., 2017; Rashkin et al., 2017). A plethora of models exist that rely on a range of data mined from various social platforms – Facebook, Twitter, Reddit – and modalities – text, images, or both. These existing models rely on different text features (*e.g.,* linguistic structure, lexical, psycholinguistic features, biased and subjective language), image features, and user interaction features (*e.g.,* engagement, network structure, temporal patterns). However, current literature lacks details for how these models transfer, *e.g.,* to out-of-domain data, across platforms, or across languages.

Moreover, there is a gap in understanding the underlying behavior of the decision-making processes behind model output. It is not clear why models for deception detection (particularly neural or deep learning based models) are making certain predictions (*e.g., why* one item is predicted as deceptive versus not). Evaluation of model performance with one metric such as accuracy or F1 score is not enough. For such complex prediction tasks like deception detection for which humans often disagree (Karduni et al., 2018; Karduni et al., 2019; Ott et al., 2011; Harris, 2012), we need more rigorous evaluation of neural model behavior and a cohesive system for comparing model results across bodies of research.

Evaluations need to explicitly measure the extent to which model performance is affected when new data is supplied, *e.g.,* data with a different topic distribution or how well-performing models on English

Proceedings of the 3rd International Workshop on Rumours and Deception in Social Media (RDSM), pages 1–13
Barcelona, Spain (Online), December 12, 2020.

data will perform on non-English inputs. There is also a critical need for evaluations that highlight when a model is correct, which examples have the ability to explain why, and, most importantly, conclude why a user should trust a given model in an interpretable manner. Arguments in favor of the above requirements to model performance are well-aligned with recent work on machine learning interpretability, trust, fairness, accountability, and reliability (Lipton, 2018; Doshi-Velez and Kim, 2017; Hohman et al., 2018). Another key argument in favor of rigorous evaluation is the lack of benchmark datasets and the need for reproducibility. Often, researchers do not make models or datasets publicly available and the details to reconstruct their models and experimental setup are not sufficient, which prevents other researchers from performing rigorous comparisons. To reproduce our key findings and experiments, we will make our framework available through interactive Jupyter notebooks available at publication.

The focus of the current work and our main contribution is an *extensive evaluation of neural model robustness across frequently encountered factors in real-world applications of digital deception models* such as new domains, languages, modalities, and upgrades to new state-of-the-art text embeddings.

2   Related Work

Deep learning systems have been adopted in many areas from medicine to autonomous driving (Ahmad et al., 2018; Claybrook and Kildare, 2018) and as these algorithms are incorporated, the need for explainable and transparent models becomes more urgent. One approach researchers use to overcome the inherent ambiguity of these black-box methods is to develop additional models to learn and explain the decisions of existing models, analyze when these models fail, and introduce a human-in-the-loop component to improve performance (Ribeiro et al., 2016; Murdoch et al., 2019; Poursabzi-Sangdeh et al., 2018). Another way to tackle this challenge is to design models with a goal of interpretability in place when development starts (Ridgeway et al., 1998; Rudin, 2018; Gilpin et al., 2018; Lahav et al., 2018; Hooker et al., 2019). Model performance beyond accuracy, precision, and recall measures are essential in order to build trustworthy and reliable models tasked with making decisions capable of significantly affecting the end users (Hohman et al., 2018; Dodge et al., 2019b; Hohman et al., 2019).

In this study, we do not focus on developing new visualization techniques to explain deception detection models using intrinsic or post-hoc explanations (Yang et al., ; Shu et al., 2019a; Reis et al., 2019; Wallace et al., 2019) or measure their interpretability (Mohseni et al., 2019). Instead, we thoroughly evaluate model performance using error analysis under several real-world conditions e.g., across domains and input types to understand reliability and the underlying behavior of decision making processes that will in turn increase model explainability and interpretability.

Deception Detection Detecting suspicious or deceptive news online is a well explored area of study. Current research efforts focus on broadly classifying between suspicious and trustworthy content (Volkova et al., 2017; Shu et al., 2020) to more specific distinctions such as between propaganda, hoax, satire, and trustworthy news (Rashkin et al., 2017), to examining the behavior of malicious users and bots (Glenski and Weninger, 2018; Kumar et al., 2017; Kumar et al., 2018), or analyzing misinformation and rumor spread patterns over time (Kwon et al., 2017; Vosoughi et al., 2018). Methods for classification vary from random forests to deep neural networks and the addition of enriched features such as images, temporal and structural attributes, and linguistic features have been shown to boost model performance over relying on textual characteristics alone (Wang, 2017; Qazvinian et al., 2011; Kwon et al., 2013).

With the high impact of digital deception on offline, real-world events, effective detection models are a critical concern. However, interpretable evaluations of model performance beyond traditional precision, recall, or F score measures are essential to building trustworthy and reliable systems when deep learning or machine learning models are tasked with making decisions capable of significantly affecting end users (Dodge et al., 2019b; Volkova et al., 2019; Hohman et al., 2019; Hohman et al., 2018).

Many studies have relied on Recurrent Neural Networks (RNNs) or variations, particularly Long Short Term Memory (LSTM) layers, for deception detection tasks (Ma et al., 2016; Chen et al., 2018; Rath et al., 2017; Zubiaga et al., 2018; Zhang et al., 2019). Others have used Convolutional Neural Networks (CNN) (Ajao et al.,), or variations of LSTM architectures such as including attention mechanisms (Guo et al., ; Li et al., 2019) which are typically dependent on specific tasks or parameter tuning of state-of-the-art deception detection models. For the purposes of consistency across our experiments, we rely on standard LSTM models that underpin much of the recent work rather than tuning each architecture by task and data or comparing the wider range of recent state-of-the-art architectures. This enables us to make more accurate comparisons of the factors related to robustness that we vary in our experiments. Developing novel models or comparing all state-of-the-art architectures is beyond the scope of this paper.

Evaluating Model Robustness Measuring robustness of a model goes beyond fortifying against intentional manipulations. Corrupted data inputs are possible in real-world scenarios without harboring the intent to fool a system and models should behave as expected in such cases, *e.g.*, simple image transformations, missing or misspelled text. Several key studies have endeavored to develop methods to evaluate model robustness (Zheng et al., 2016; Hein and Andriushchenko, 2017; Liu et al., 2018; Dodge et al., 2019a). In this study we focus on evaluating model robustness to out-of-domain data, multilingual data, and multimodal inputs combined with conceptually different text embedding techniques to gain deeper insights into how much "learning and understanding" neural network models actually have. Different studies have separately examined the relationships between extracted features across modalities, across languages, and across domains (Zhou et al., 2020; Capuozzo et al., 2020; Zotova et al., 2020). To the best of our knowledge, this extensive evaluation of deception detection model robustness across multiple tasks has not been presented before in one unified work.

3 Methodology

In this section we discuss the state-of-the-art neural models for deception detection used in this paper and describe our approach for evaluating model robustness across domains, modalities, and languages, a summary of which is presented in Table 1.

3.1 Neural Models for Deception Detection

After an extensive analysis of published neural deception classification models, we adopt similar architectures that consist of a commonly used RNN architecture, LSTM layers, and rely on combinations of text, linguistic cue (LC), and image vectors for our experiments. Lexical vectors and linguistic cues include those often used for classification, notably encoding for biased and otherwise subjective language (Rashkin et al., 2017; Shu et al., 2019b). We employ LIWC (Pennebaker et al., 2001) and several lexical dictionaries (assertive verbs, hedges, factives, implicatives, etc. (Recasens et al., 2013)) to construct frequency vectors for model input. Image vectors are representations extracted from the last layer of the state-of-the-art ResNet architecture (He et al., 2016). We focus our analysis on both models with single modalities *e.g.*, text and multimodal models *e.g.*, text and image, as well as classification tasks over a varied number of classes. We create 80/20 splits in each dataset for train and test sets.

We implement the binary classification model over trustworthy and deceptive samples using text and lexical input features to evaluate the efficacy and robustness of testing on out-of-domain data. We additionally implement the binary (trustworthy versus deceptive), the 3-way (trustworthy, propaganda, disinformation), and the 4-way (clickbait, satire, hoax, conspiracy) classification models with the enhancement of image input features for our evaluation of multimodal models. Our final model to assess the robustness of these deep learning approaches is the 3-way classification of clickbait, conspiracy, and

Model	Platform(s)	Input	Text Embeddings	Detection Task (Classes)	Robustness Task
M_1	Twitter Reddit	Text + LC	BERT	trustworthy, deceptive	Cross-Domain
M_2	Twitter	Text + Image + LC	BERT, GLoVe, ELMo	trustworthy, deceptive	Cross-Modality
M_3	Twitter	Text + Image + LC	BERT, GLoVe, ELMo	trustworthy, propaganda, disinformation	Cross-Modality
M_4	Twitter	Text + Image + LC	BERT, GLoVe, ELMo	clickbait, satire, hoax, conspiracy	Cross-Modality
M_5	Twitter	Text	Character	clickbait, conspiracy, propaganda	Cross-Language

Table 1: Overview of the deception detection models (M_1, M_2, M_3, M_4, and M_5) used for each task.

	Train		**Test**	
Data	*Trustworthy*	*Deceptive*	*Trustworthy*	*Deceptive*
Twitter	14.9k	28.0k	3.7k	7.2k
Reddit	21.6k	21.5k	5.4k	5.5k
Twitter + Reddit	36.6k	49.9k	–	–

Table 2: Distributions across classes within train and test data used in the cross-domain analyses (M_1).

propaganda news sources across multiple languages. We employ similar neural network architectures for all models, which consists of a pre-trained text embedding layer followed by an LSTM layer and a single dense layer. The output from this sub-network is concatenated to the image vectors which have been transformed by two dense layers. In the architectures of our multi-modal models, our joint text and image representations are concatenated to the lexical cues vector as input to the final two fully connected layers in the model. All layers use dropout.[1]

In contrast to the Cross-Domain and Cross-Modality models, the multilingual Cross-Language models do not incorporate linguistic cues or other lexical features because of the inconsistency across multiple languages – some of the linguistic features and lexicons are not available for all languages (*e.g.*, bias and subjective language dictionaries). The multilingual model (M_5) architecture is constructed by the pre-trained character-level text embedding layer followed by two convolution layers, then a max pooling layer, and finally two dense, fully connected layers.

3.2   Datasets

We use a previously released, public list of news sources[2] annotated along a spectrum of deceptive – clickbait, hoax, satire, conspiracy, propaganda, or disinformation – and verified news sources who typically spread factual content (which we denote as "trustworthy" in our experiments). These annotated news sources focused on activity in 2016, therefore, we collected the following datasets for the same 12 month period of activity. Note, there are limitations when annotations are done on the sources level, however, similar to related work (Vosoughi et al., 2018; Lazer et al., 2018; Glenski et al., 2018) we advocate focusing on the news sources rather than individual stories because we view the definitive element of deception to be the intent and the tactics of the news source.

In our ***Cross-Domain*** analyses, we consider two domains – Twitter and Reddit. The Twitter component of the M_1 dataset is comprised of English retweets from official twitter accounts for news media described above, containing only text-based posts. The Reddit component comprises all top-level comments in response to posts on Reddit in 2016 where the post contained a link to a web domain associated with one of the news accounts of interest. We create a binary classification dataset, analogous to the Twitter dataset used for M_2, that collapses news source annotations to either trustworthy or deceptive and down-samples the Reddit comments to have approximately the same volume of content ($N = 54k$). This allows the joint *Twitter + Reddit* dataset to be balanced between the two domains for cross-domain analyses, as shown in Table 2 where we present the class distributions for the M_1 dataset.

Our multimodal Twitter data, for the ***Cross-Modality*** analyses, is comprised of the same collection of English retweets from annotated news media's official accounts used to build the Twitter component of the M_1 dataset, filtered to those tweets that include images. Each retweet comprises a unique image and body of text. We create datasets for three classification tasks: (1) propaganda, disinformation, or trustworthy (M_3; $N = 54.4k$), (2) clickbait, conspiracy, hoax, or satire (M_4; $N = 2.5k$), and (3) a binary classification dataset (M_2; $N = 54.4k$) of trustworthy versus deceptive tweets in the dataset used for M_3 by collapsing the sub-classes of propaganda and disinformation into a single deceptive class. The

---

[1]For each model, we perform a random hyper-parameter tuning search independently consisting of ten different configurations of model hyper-parameters. The batch size, the number of training epochs, drop out rate, and the recurrent layer dimensionality as well as the optimizer and learning rate are among the tunable parameters. Every model had a unique set of final hyper-parameters, however, the most common configurations consisted of the Adam optimizer with a learning rate of 0.0001, a drop out rate between 0.2 and 0.25, and 10 epochs of training.

[2]We use the following publicly available news source annotations for *deceptive news sources*: `https://www.cs.jhu.edu/~svitlana/data/SuspiciousNewsAccountList.tsv` and *trustworthy news sources*: `https://www.cs.jhu.edu/~svitlana/data/VerifiedNewsAccountList.tsv`

breakdown of these three tasks is presented in Table 1.

In contrast to the previous, English-only datasets, our multilingual Twitter dataset for the ***Cross-Language*** analyses comprises 7,316 tweets across five languages (English, French, German, Russian, and Spanish) from a similar collection of tweets as described above for English Twitter without the restriction to only English-text. Because of an over-representation of English-tweets in the original collection, we sample 1,500 examples for each language evenly distributed across the three classes, *i.e.,* 500 clickbait samples, 500 conspiracy samples, and 500 propaganda samples. For the least represented language, Russian, we use all available tweets: 478 clickbait samples, 338 conspiracy samples, and 500 propaganda samples. For all languages, we allocate 80% to train and 20% to test.

3.3 Evaluating Neural Model Robustness

We investigate the robustness of digital deception models when evaluated across three dimensions: domains, modalities and languages, seeking to answer three key research questions. Table 1 outlines the model architectures, inputs, datasets, and prediction tasks for all robustness analyses.

In our ***Cross-Domain*** analyses, we evaluate model robustness across two popular social platforms: Twitter and Reddit. We compare model performance on a binary classification task (M_1 in Table 1) using the binary English Twitter and Reddit datasets summarized in Table 2. We train three LSTM models using textual and lexical input features – (1) trained on Twitter content only, (2) trained on Reddit content only, and (3) trained on both the Twitter and Reddit content. All three models rely on pre-trained BERT word embeddings that are fine-tuned during training and use the same neural architecture and hyper-parameters described in section 3.1. To account for inherent platform inconsistencies between Twitter and Reddit, we limit the input text length to 100 words and use a common vocabulary between the three model setups. We evaluate each model on two held out test sets composed of held out examples from Twitter and Reddit allowing us to test performance within the same domain (*e.g.,* train and test on Twitter data) and on out-of-domain data (*e.g.,* train on Twitter and test on Reddit).

In our ***Cross-Modality*** analyses, we evaluate how multiple modalities can influence the predictions of a model. We train three LSTM classifiers to predict falsified news at different levels of granularity. The first model (M_2) is trained to distinguish between trustworthy and deceptive news using text, lexical cues, and extracted image features. The second model (M_3) uses more fine-grained labels of digital deception to further separate deceptive news into propaganda and disinformation. Finally, (M_4) is a 4-way classification task to predict tweets from four types of deceptive news types: clickbait, conspiracy, hoax, and satire. Each classifier is trained three separate times using a different text embedding strategy that allows for fine-tuning during training: BERT, ELMo, or GloVe. We use the Tensorflow Object Detection API (Huang et al., 2017) to extract the primary COCO[3] (Common Objects in COntext) objects from the images of the tweets. In this analysis, we also seek to answer, in particular, the question: *Which errors do text embeddings or detected objects contribute to in the multimodal setup?*

In our ***Cross-Language*** analyses, we evaluate changes in model performance when a convolutional neural network (M_5) developed to distinguish fine-grained differences in digital deception (*clickbait* versus *conspiracy* versus *propaganda*) is trained and evaluated across multiple languages. We examine the impacts on performance when the model is trained in the context of multiple languages (*i.e.,* a single model trained using an aggregated dataset of English, Russian, German, Spanish, and French samples) and evaluated on a single language as well as when the model is trained in the context of a single language and tested on the same language using 5-fold cross-validation.

4 Experimental Results

In this section, we highlight the results of our experiments regarding model robustness on out-of-domain data (Cross-Domain analyses), across multimodal inputs using various text embeddings (Cross-Modality analyses), and multilingual inputs (Cross-Language analyses).

---

[3]http://cocodataset.org/#home

"RT fluoridated water now linked to diabetes & lowered IQ still drinking it"

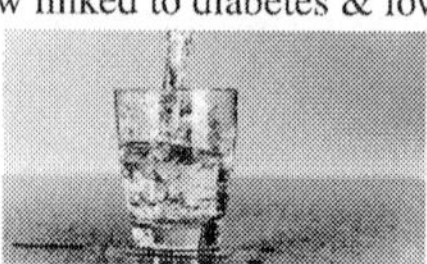

Figure 1: Incorrect classification with high confidence example: a tweet annotated *Conspiracy* classified as *Clickbait* with 92.6% confidence by the M_4 (ELMo embeddings) model.

	True Class	Test on **Twitter**: Predicted Trustworthy	Test on **Twitter**: Predicted Deceptive	Test on **Reddit**: Predicted Trustworthy	Test on **Reddit**: Predicted Deceptive
Train on **Twitter**	Trustworthy	46% (1.72k)	54% (2.02k)	89% (4.77k)	11% (573)
	Deceptive	19% (1.38k)	81% (5.79k)	90% (4.72k)	10% (517)
Train on **Reddit**	Trustworthy	87% (3.27k)	13% (469)	62% (3.28k)	38% (2.06k)
	Deceptive	89% (6.41k)	11% (757)	51% (2.69k)	49% (2.55k)
Train on **Twitter + Reddit**	Trustworthy	60% (2.26k)	40% (1.49k)	59% (3.22k)	41% (2.21k)
	Deceptive	20% (1.43k)	80% (5.75k)	28% (1.55k)	72% (3.95k)

Figure 2: Confusion matrices for cross-domain analyses using models that classify content as Trustworthy or Deceptive. Cells are shaded by the rate of (mis)classification; Correct predictions are shaded in blue, misclassifications in red.

Before we do so, we manually validate the models qualitatively by examining incorrectly classified tweets with high confidences. For example, the 4-way classification model, M_4, that relies on pre-trained ELMo embeddings shows the best performance, but makes errors that resemble human mistakes. The tweet shown in Figure 1 is an example of a high-confidence misclassification that could intuitively be similarly made by a human annotator. From the text or image alone the true classification is not immediately clear or intuitive from a human standpoint. The text implies conspiracy, but it is not obvious from the image – a stock image of a glass of water being poured – to which class it should belong. This analysis of high-confidence incorrect classifications provides a better understanding of how high confidence misclassifications (especially high-confidence which can serve as a proxy for where end-users *should* – or would feel that they should – trust the model) occur and under what conditions.

4.1 Cross-Domain Performance

We present the confusion matrices for the cross-domain experiments in Figure 2 and, as expected, we observe the best individual performance when models are trained and tested on same-domain (same-platform) data. Additionally, we see that performance suffers on models tested on out-of-domain data (models trained on Twitter but test on Reddit and vice-versa). However, we find that both of the cross-domain models (M_1) trained on either the Twitter or Reddit data alone over-predict the *trustworthy* class when tested on out-of-domain test data. This is irrespective of the distribution of class instances (shown in Table 2) – whether it is imbalanced (as it is in Twitter) or not (as in Reddit). As illustrated in the final row of confusion matrices, we see that the best overall performance is achieved by the model with *domain diverse* training data – trained on both the Twitter and Reddit training datasets combined. This model *performs comparably on both the Twitter and Reddit held out test sets and as well or better than the individually trained models when tested on in-domain samples.*

Our findings confirm that data from new domains (*i.e.,* social platforms) has a high impact on model performance even in the cases where classes are defined in the same way (as is the case with our Twitter and Reddit datasets that rely on annotations of the same pool of news sources as trustworthy or deceptive).

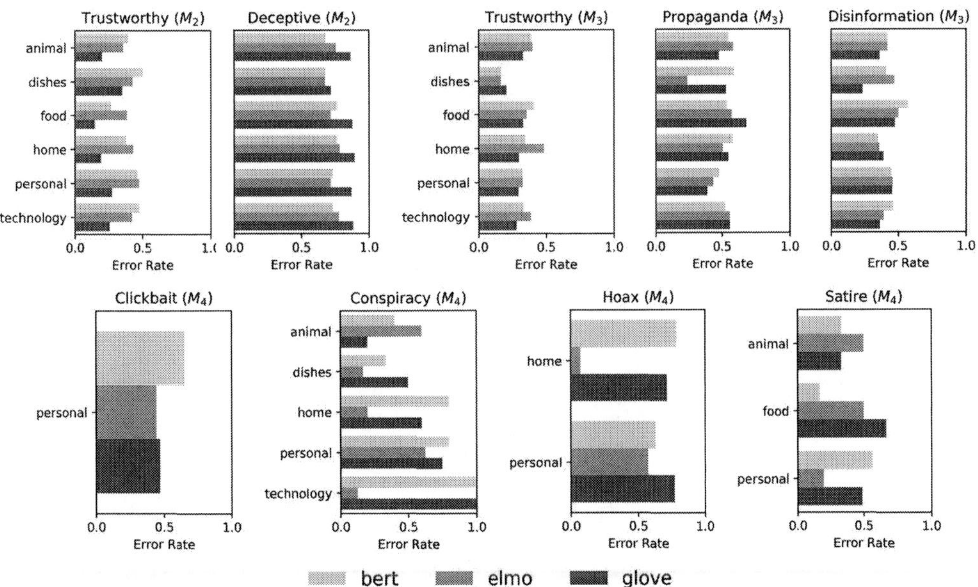

Figure 3: Error rates for Cross-Modality models for each of the six categories of primary COCO objects detected where there were at least five true class examples for which to calculate error rates.

On the Twitter test data, the domain-diverse model is even able to more accurately identify trustworthy content (60% true positives compared to 46% when trained on Twitter data alone) and reduce the number of false positive predictions of Deceptive content (40% of trustworthy tweets classified as deceptive compared to 54%). This indicates that even if used only for a single platform, engaging diverse training data examples across multiple platforms can result in a more robust, more trustworthy model.

4.2 Cross-Modality Performance

Next, we examine the performance of our multi-modal models, presented in Figure 3, across each of the three tasks and two modalities: different state-of-the-art text embeddings used by the model to represent the text component of the news post and the most distinguishable objects present in the image component of the news post to be classified. We observe several significant findings related to these modality-specific features. Because of the numerous COCO (Lin et al., 2014) classes, we group similar objects into super-categories[4]. For example, the *personal* category contains people, umbrellas, backpacks, etc.

When we examine the performance of the binary cross-modality (M_2) model, we see the best model performance on trustworthy tweets – with about 50% fewer misclassifications than on deceptive tweets – consistently across every object class and embedding type. In particular, the model that utilizes the pre-trained GloVe embeddings achieves the lowest error rate of 19.35%. We also find that the model using the GloVe embeddings has the best performance — the lowest error rate — among the 3-way (M_3) models with an error rate of 23.53%. Figure 4 illustrates an example tweet where the GloVe-based (M_3) model has correctly classified the tweet as disinformation but the models that rely on the BERT and ELMo text embeddings to represent the text component of the tweet incorrectly classify it as a propaganda tweet.

We find that Disinformation tweets where images contain food items have a misclassification rate greater than 50% for 3-way classification models (M_3) – the highest of all objects. In contrast, trustworthy tweets have the lowest misclassification error of any class across the model types for all objects at 32.15%. However, one anomaly is the rate of misclassification (48.19%) of trustworthy tweets with home-related objects (such as books) from the ELMo (M_3) model. An example of such a tweet where this ELMo model is incorrect while the others are correct is shown in Figure 5. In all three classes, we see the lowest misclassification error when kitchen-related objects are present in the image. In particular, propaganda tweets, where kitchen and personal objects appear in the image, are correctly classified

---

[4]https://tech.amikelive.com/node-718/what-object-categories-labels-are-in-coco-dataset/

Figure 4: An example of a disinformation tweet that the GloVe (M_3) model is able to correctly identify (55.17% confidence) but the BERT and ELMo (M_3) models misclassify as propaganda (confidences of 48.29% and 60.86%).

Figure 5: The text and image from a trustworthy tweet. The ELMo model misclassifies (as disinformation with 85.74% confidence) while the BERT and GloVe models correctly classify it as trustworthy (with confidences of 76.55% and 44.27%).

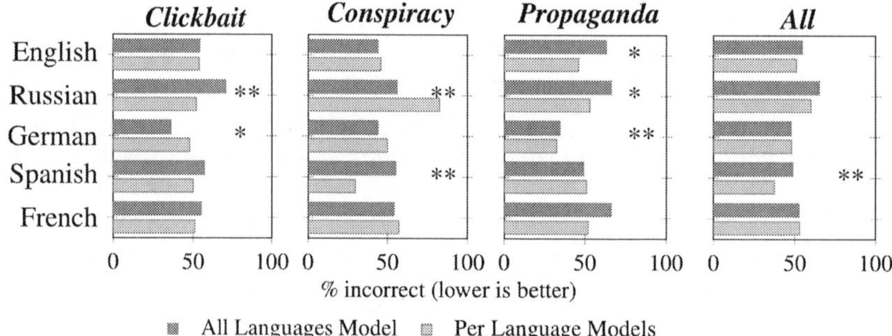

Figure 6: Bar plots display the number of incorrect predictions for each class (Clickbait, Conspiracy, Propaganda) and overall (All) for the multilingual (All Languages) model and single-language (Per Language) models over the five languages: English, Russian, German, Spanish, and French. Statistically significant differences in performance are indicated with * for $p < 0.05$ and ** for $p < 0.01$ (MWU).

more often than misclassified (an error rate less than 45%). Interestingly, the misclassification rate is similar regardless of the embedding type (between 16.67% and 20.08%) for trustworthy tweets with kitchen objects. For propaganda tweets with kitchen objects, ELMo embeddings produce the lowest misclassification rate (23.52%).

For the 4-way classification (M_4) task, we find that conspiracy tweets containing images with kitchen objects have the lowest error rate at 33.33% while conspiracy tweets with technology have the highest error rate at 70.83% compared to the remaining three deceptive classes. Predictions with ELMo embeddings have the lowest misclassification rate (36.13%) across all four classes and all object types compared to BERT (67.58%) and GloVe (61.09%). We see this particularly in cases where the BERT and GloVe models have high error rates, *e.g.*, conspiracy tweets that contain technology-related objects and hoax tweets that include home objects.

4.3   Cross-Language Performance

Finally, we illustrate the performance of the six cross-languages models in Figure 6 (in these plots, lower bars indicated higher performance): a single multilingual model trained jointly on multiple languages

(*All Languages Model*) and five models trained on each of the languages individually (*Per Language Models*, one for each of the five languages). As shown in Figure 6, the multilingual model trained jointly on multiple languages produces incorrect classifications most often on tweets that were labeled as propaganda when tested on English data while both the multilingual and single-language models tested on English data experience a steady number of errors over clickbait and conspiracy.

When we compare the jointly trained 'All Languages' model versus individually trained 'Per-Language' models across the other languages, we see that the single-language ('Per Language') models trained separately with English, Russian, German, and Spanish outperform the aggregated-language model tested on the same languages with regards to propaganda. The single-language model trained and tested on Spanish data displays the best performance on conspiracy. Analytically, we can see the difference in model performance when looking at each class individually and over all classes collectively. The aggregated-language model tested with German data shows better performance over clickbait and conspiracy while making more errors with propaganda. However, the single-language model trained and tested on German data sees lesser performance over clickbait and conspiracy while making fewer errors with propaganda. When we review the aggregated- and single-language German models over all classes, they display a similar performance.

5 Discussion and Future Work

We have presented an extensive evaluation of the robustness of digital deception models across frequently encountered real-world scenarios that would be necessary to benchmark against to identify reasonable performance for models considered for deployment. Our analyses have identified several trends in behavior across multiple robustness tasks and granularities of deception detection tasks (from binary to 4-way classification). To the best of our knowledge, we are the first to present this type of evaluation of deception detection model robustness.

In our robustness analyses detailed above, we have illustrated the danger of relying on single performance measurement, metrics, or analyses by showcasing how a model achieving optimal performance or significant performance increases on a specific task, domain, or context (*e.g.,* multilingual, multimodal) may significantly under-perform with slight alterations in scope or context and other frequently encountered factors in real-world applications. Further, we have shown several ways that out-of-domain, multilingual, and multiple modality inputs affect model performance and potential methods to mitigate the impact. For example, the out-of-domain analyses showed that a domain-diverse set of training examples led to higher performance on both domains in held out test sets compared to models trained on domain-specific examples. Additionally, the cross-modality analyses illustrated how frequently human-like classification mistakes can be made and which areas of the data distribution are not well represented, *e.g.,* clickbait tweets with images containing non-personal related objects. Results highlighted in the current work open several avenues of future work to develop, evaluate, and understand neural deception detection models in the context of real-world applications.

Future work will investigate the performance of a variety of such domain diverse training data compared to domain-specific models to identify if this hypothesis holds or if there may be other confounding factors related to specific platforms. In the current work, we have focused on quantifying the performance using a simplified, shared architecture that has been frequently used in deception detection models. However, future work can leverage our interactive Jupyter notebooks that will be made available at publication for reproducible extensions using our analysis framework to benchmark the performance of more complex, or newly developed state-of-the-art neural architectures.

Acknowledgements

This research was supported by the Laboratory Directed Research and Development Program at Pacific Northwest National Laboratory, a multiprogram national laboratory operated by Battelle for the U.S. Department of Energy.

References

Muhammad Aurangzeb Ahmad, Carly Eckert, and Ankur Teredesai. 2018. Interpretable machine learning in healthcare. In *Proceedings of the 2018 ACM International Conference on Bioinformatics, Computational Biology, and Health Informatics*, pages 559–560. ACM.

Oluwaseun Ajao, Deepayan Bhowmik, and Shahrzad Zargari. Fake news identification on twitter with hybrid cnn and rnn models. In *Proceedings of the 9th International Conference on Social Media and Society*.

Pasquale Capuozzo, Ivano Lauriola, Carlo Strapparava, Fabio Aiolli, and Giuseppe Sartori. 2020. Decop: A multilingual and multi-domain corpus for detecting deception in typed text. In *Proceedings of The 12th Language Resources and Evaluation Conference*, pages 1423–1430.

Weiling Chen, Yan Zhang, Chai Kiat Yeo, Chiew Tong Lau, and Bu Sung Lee. 2018. Unsupervised rumor detection based on users' behaviors using neural networks. *Pattern Recognition Letters*, 105:226–233.

Joan Claybrook and Shaun Kildare. 2018. Autonomous vehicles: No driver... no regulation? *Science*, 361(6397):36–37.

Jesse Dodge, Suchin Gururangan, Dallas Card, Roy Schwartz, and Noah Smith. 2019a. Show your work: Improved reporting of experimental results. *arXiv preprint arXiv:1909.03004*.

Jonathan Dodge, Q Vera Liao, Yunfeng Zhang, Rachel KE Bellamy, and Casey Dugan. 2019b. Explaining models: an empirical study of how explanations impact fairness judgment. In *Proceedings of the 24th International Conference on Intelligent User Interfaces*, pages 275–285. ACM.

Finale Doshi-Velez and Been Kim. 2017. Towards a rigorous science of interpretable machine learning. *arXiv preprint arXiv:1702.08608*.

Leilani H Gilpin, David Bau, Ben Z Yuan, Ayesha Bajwa, Michael Specter, and Lalana Kagal. 2018. Explaining explanations: An overview of interpretability of machine learning. In *2018 IEEE 5th International Conference on data science and advanced analytics (DSAA)*, pages 80–89. IEEE.

Maria Glenski and Tim Weninger. 2018. How humans versus bots react to deceptive and trusted news sources: A case study of active users. In *Proceedings of the IEEE/ACM International Conference on Advances in Social Networks Analysis and Mining (ASONAM)*. IEEE/ACM.

Maria Glenski, Tim Weninger, and Svitlana Volkova. 2018. Propagation from deceptive news sources who shares, how much, how evenly, and how quickly? *IEEE Transactions on Computational Social Systems*, 5(4):1071–1082.

Han Guo, Juan Cao, Yazi Zhang, Junbo Guo, and Jintao Li. Rumor detection with hierarchical social attention network. In *Proceedings of the 27th ACM International Conference on Information and Knowledge Management*.

Christopher Glenn Harris. 2012. Detecting deceptive opinion spam using human computation. In *Workshops at the Twenty-Sixth AAAI Conference on Artificial Intelligence*.

Kaiming He, Xiangyu Zhang, Shaoqing Ren, and Jian Sun. 2016. Deep residual learning for image recognition. In *Proceedings of the IEEE conference on computer vision and pattern recognition*, pages 770–778.

Matthias Hein and Maksym Andriushchenko. 2017. Formal guarantees on the robustness of a classifier against adversarial manipulation. In *Advances in Neural Information Processing Systems*, pages 2266–2276.

Fred Matthew Hohman, Minsuk Kahng, Robert Pienta, and Duen Horng Chau. 2018. Visual analytics in deep learning: An interrogative survey for the next frontiers. *IEEE transactions on visualization and computer graphics*.

Fred Hohman, Andrew Head, Rich Caruana, Robert DeLine, and Steven M Drucker. 2019. Gamut: A design probe to understand how data scientists understand machine learning models. In *Proceedings of the 2019 CHI Conference on Human Factors in Computing Systems*, page 579. ACM.

Sara Hooker, Dumitru Erhan, Pieter-Jan Kindermans, and Been Kim. 2019. A benchmark for interpretability methods in deep neural networks. In *Advances in Neural Information Processing Systems*, pages 9734–9745.

Jonathan Huang, Vivek Rathod, Chen Sun, Menglong Zhu, Anoop Korattikara, Alireza Fathi, Ian Fischer, Zbigniew Wojna, Yang Song, Sergio Guadarrama, et al. 2017. Speed/accuracy trade-offs for modern convolutional object detectors. In *Proceedings of the IEEE conference on computer vision and pattern recognition*.

Georgi Karadzhov, Pepa Gencheva, Preslav Nakov, and Ivan Koychev. 2017. We built a fake news & click-bait filter: What happened next will blow your mind! In *Proceedings of the International Conference on Recent Advances in Natural Language Processing*.

Alireza Karduni, Ryan Wesslen, Sashank Santhanam, Isaac Cho, Svitlana Volkova, Dustin Arendt, Samira Shaikh, and Wenwen Dou. 2018. Can you verifi this? studying uncertainty and decision-making about misinformation using visual analytics. In *Twelfth international AAAI conference on web and social media*.

Alireza Karduni, Isaac Cho, Ryan Wesslen, Sashank Santhanam, Svitlana Volkova, Dustin L Arendt, Samira Shaikh, and Wenwen Dou. 2019. Vulnerable to misinformation?: Verifi! In *Proceedings of the 24th International Conference on Intelligent User Interfaces*, pages 312–323. ACM.

Srijan Kumar, Justin Cheng, Jure Leskovec, and VS Subrahmanian. 2017. An army of me: Sockpuppets in online discussion communities. In *Proceedings of the 26th International Conference on World Wide Web*, pages 857–866.

Srijan Kumar, Bryan Hooi, Disha Makhija, Mohit Kumar, Christos Faloutsos, and VS Subrahmanian. 2018. Rev2: Fraudulent user prediction in rating platforms. In *Proceedings of the Eleventh ACM International Conference on Web Search and Data Mining*, pages 333–341. ACM.

Sejeong Kwon, Meeyoung Cha, Kyomin Jung, Wei Chen, and Yajun Wang. 2013. Prominent features of rumor propagation in online social media. In *Proceedings of the 13th International Conference on Data Mining*, pages 1103–1108. IEEE.

Sejeong Kwon, Meeyoung Cha, and Kyomin Jung. 2017. Rumor detection over varying time windows. *PloS One*, 12(1):e0168344.

Owen Lahav, Nicholas Mastronarde, and Mihaela van der Schaar. 2018. What is interpretable? using machine learning to design interpretable decision-support systems. *arXiv preprint arXiv:1811.10799*.

David MJ Lazer, Matthew A Baum, Yochai Benkler, Adam J Berinsky, Kelly M Greenhill, Filippo Menczer, Miriam J Metzger, Brendan Nyhan, Gordon Pennycook, David Rothschild, et al. 2018. The science of fake news. *Science*, 359(6380):1094–1096.

Quanzhi Li, Qiong Zhang, and Luo Si. 2019. Rumor detection by exploiting user credibility information, attention and multi-task learning. In *Proceedings of the 57th Annual Meeting of the Association for Computational Linguistics*, pages 1173–1179.

Tsung-Yi Lin, Michael Maire, Serge Belongie, James Hays, Pietro Perona, Deva Ramanan, Piotr Dollár, and C Lawrence Zitnick. 2014. Microsoft coco: Common objects in context. In *European conference on computer vision*, pages 740–755. Springer.

Zachary C Lipton. 2018. The mythos of model interpretability. *Queue*, 16(3):31–57.

Xuanqing Liu, Minhao Cheng, Huan Zhang, and Cho-Jui Hsieh. 2018. Towards robust neural networks via random self-ensemble. In *Proceedings of the European Conference on Computer Vision (ECCV)*, pages 369–385.

Jing Ma, Wei Gao, Prasenjit Mitra, Sejeong Kwon, Bernard J Jansen, Kam-Fai Wong, and Meeyoung Cha. 2016. Detecting rumors from microblogs with recurrent neural networks. In *Ijcai*, pages 3818–3824.

Tanushree Mitra, Graham P Wright, and Eric Gilbert. 2017. A parsimonious language model of social media credibility across disparate events. In *Proceedings of the ACM Conference on Computer Supported Cooperative Work and Social Computing (CSCW)*, pages 126–145. ACM.

Sina Mohseni, Eric Ragan, and Xia Hu. 2019. Open issues in combating fake news: Interpretability as an opportunity. *arXiv preprint arXiv:1904.03016*.

W James Murdoch, Chandan Singh, Karl Kumbier, Reza Abbasi-Asl, and Bin Yu. 2019. Interpretable machine learning: definitions, methods, and applications. *arXiv preprint arXiv:1901.04592*.

Myle Ott, Yejin Choi, Claire Cardie, and Jeffrey T Hancock. 2011. Finding deceptive opinion spam by any stretch of the imagination. In *Proceedings of the 49th annual meeting of the association for computational linguistics: Human language technologies-volume 1*, pages 309–319. Association for Computational Linguistics.

James W Pennebaker, Martha E Francis, and Roger J Booth. 2001. Linguistic inquiry and word count: Liwc 2001. *Mahway: Lawrence Erlbaum Associates*, 71.

Forough Poursabzi-Sangdeh, Daniel G Goldstein, Jake M Hofman, Jennifer Wortman Vaughan, and Hanna Wallach. 2018. Manipulating and measuring model interpretability. *arXiv preprint arXiv:1802.07810*.

Vahed Qazvinian, Emily Rosengren, Dragomir R Radev, and Qiaozhu Mei. 2011. Rumor has it: Identifying misinformation in microblogs. In *Proceedings of the Conference on Empirical Methods in Natural Language Processing (EMNLP)*, pages 1589–1599.

Hannah Rashkin, Eunsol Choi, Jin Yea Jang, Svitlana Volkova, and Yejin Choi. 2017. Truth of varying shades: Analyzing language in fake news and political fact-checking. In *Proceedings of the 2017 Conference on Empirical Methods in Natural Language Processing*, pages 2921–2927.

Bhavtosh Rath, Wei Gao, Jing Ma, and Jaideep Srivastava. 2017. From retweet to believability: Utilizing trust to identify rumor spreaders on twitter. *Proceedings of the IEEE/ACM International Conference on Advances in Social Networks Analysis and Mining (ASONAM)*.

Marta Recasens, Cristian Danescu-Niculescu-Mizil, and Dan Jurafsky. 2013. Linguistic models for analyzing and detecting biased language. In *Proceedings of the 51st Annual Meeting of the Association for Computational Linguistics (Volume 1: Long Papers)*, pages 1650–1659.

Julio Reis, André Correia, Fabrício Murai, Adriano Veloso, and Fabrício Benevenuto. 2019. Explainable machine learning for fake news detection. In *Proceedings of the 10th ACM Conference on Web Science*, pages 17–26. ACM.

Marco Tulio Ribeiro, Sameer Singh, and Carlos Guestrin. 2016. Why should i trust you?: Explaining the predictions of any classifier. In *Proceedings of the 22nd ACM SIGKDD international conference on knowledge discovery and data mining*, pages 1135–1144. ACM.

Greg Ridgeway, David Madigan, Thomas Richardson, and John O'Kane. 1998. Interpretable boosted naïve bayes classification. In *KDD*, pages 101–104.

Victoria L Rubin, Niall J Conroy, Yimin Chen, and Sarah Cornwell. 2016. Fake news or truth? Using satirical cues to detect potentially misleading news. In *Proceedings of NAACL-HLT*, pages 7–17.

Cynthia Rudin. 2018. Please stop explaining black box models for high stakes decisions. *arXiv preprint arXiv:1811.10154*.

Kai Shu, Amy Sliva, Suhang Wang, Jiliang Tang, and Huan Liu. 2017. Fake news detection on social media: A data mining perspective. *ACM SIGKDD Explorations Newsletter*, 19(1):22–36.

Kai Shu, Limeng Cui, Suhang Wang, Dongwon Lee, and Huan Liu. 2019a. defend: Explainable fake news detection. In *Proceedings of the 25th ACM SIGKDD International Conference on Knowledge Discovery & Data Mining*.

Kai Shu, Suhang Wang, and Huan Liu. 2019b. Beyond news contents: The role of social context for fake news detection. In *Proceedings of the Twelfth ACM International Conference on Web Search and Data Mining*, pages 312–320. ACM.

Kai Shu, Deepak Mahudeswaran, Suhang Wang, and Huan Liu. 2020. Hierarchical propagation networks for fake news detection: Investigation and exploitation. In *Proceedings of the International AAAI Conference on Web and Social Media*, volume 14, pages 626–637.

Svitlana Volkova, Kyle Shaffer, Jin Yea Jang, and Nathan Hodas. 2017. Separating facts from fiction: Linguistic models to classify suspicious and trusted news posts on twitter. In *Proceedings of the 55th Annual Meeting of the Association for Computational Linguistics*, volume 2, pages 647–653.

Svitlana Volkova, Ellyn Ayton, Dustin L Arendt, Zhuanyi Huang, and Brian Hutchinson. 2019. Explaining multimodal deceptive news prediction models. In *Proceedings of the International AAAI Conference on Web and Social Media*, volume 13, pages 659–662.

Soroush Vosoughi, Deb Roy, and Sinan Aral. 2018. The spread of true and false news online. *Science*, 359(6380):1146–1151.

Eric Wallace, Jens Tuyls, Junlin Wang, Sanjay Subramanian, Matt Gardner, and Sameer Singh. 2019. Allennlp interpret: A framework for explaining predictions of nlp models. *arXiv preprint arXiv:1909.09251*.

William Yang Wang. 2017. "liar, liar pants on fire": A new benchmark dataset for fake news detection. *arXiv preprint arXiv:1705.00648*.

Fan Yang, Shiva K Pentyala, Sina Mohseni, Mengnan Du, Hao Yuan, Rhema Linder, Eric D Ragan, Shuiwang Ji, and Xia Ben Hu. Xfake: Explainable fake news detector with visualizations. In *The World Wide Web Conference*.

Qiang Zhang, Aldo Lipani, Shangsong Liang, and Emine Yilmaz. 2019. Reply-aided detection of misinformation via bayesian deep learning. In *WWW*, pages 2333–2343. ACM.

Stephan Zheng, Yang Song, Thomas Leung, and Ian Goodfellow. 2016. Improving the robustness of deep neural networks via stability training. In *Proceedings of the ieee conference on computer vision and pattern recognition*.

Xinyi Zhou, Jindi Wu, and Reza Zafarani. 2020. Safe: Similarity-aware multi-modal fake news detection. *arXiv preprint arXiv:2003.04981*.

Elena Zotova, Rodrigo Agerri, Manuel Núñez, and German Rigau. 2020. Multilingual stance detection in tweets: The catalonia independence corpus. In *Proceedings of The 12th Language Resources and Evaluation Conference*, pages 1368–1375.

Arkaitz Zubiaga, Elena Kochkina, Maria Liakata, Rob Procter, Michal Lukasik, Kalina Bontcheva, Trevor Cohn, and Isabelle Augenstein. 2018. Discourse-aware rumour stance classification in social media using sequential classifiers. *Information Processing & Management*, 54(2):273–290.

A Language-Based Approach to Fake News Detection Through Interpretable Features and BRNN

Yu Qiao[1], Daniel Wiechmann[2], Elma Kerz[1*]
RWTH Aachen University[1], University of Amsterdam[2]
yu.qiao@rwth-aachen.de, d.wiechmann@uva.nl
elma.kerz@ifaar.rwth-aachen.de

Abstract

'Fake news' – succinctly defined as false or misleading information masquerading as legitimate news – is a ubiquitous phenomenon and its dissemination weakens the fact-based reporting of the established news industry, making it harder for political actors, authorities, media and citizens to obtain a reliable picture. State-of-the art language-based approaches to fake news detection that reach high classification accuracy typically rely on black box models based on word embeddings. At the same time, there are increasing calls for moving away from black-box models towards white-box (explainable) models for critical industries such as healthcare, finances, military and news industry. In this paper we performed a series of experiments where bi-directional recurrent neural network classification models were trained on interpretable features derived from multi-disciplinary integrated approaches to language. We apply our approach to two benchmark datasets. We demonstrate that our approach is promising as it achieves similar results on these two datasets as the best performing black box models reported in the literature. In a second step we report on ablation experiments geared towards assessing the relative importance of the human-interpretable features in distinguishing fake news from real news.

1 Introduction

The topic of 'disinformation' – an umbrella term used to encompass a wide range of types of information disorder, "including 'fake news', rumors, deliberately factually incorrect information, inadvertently factually incorrect information, politically slanted information, and 'hyperpartisan' news" (Tucker et al., 2018) – is attracting more and more attention. This reflects a deeper concern that the prevalence of disinformation leads to an increased political polarization, decreases trust in public institutions, and undermines democracy. For example, the spread of 'fake news' – concisely defined as intentionally false information masquerading as genuine news – for financial and political gains had a potential impact on the contentious Brexit referendum or 2016 U.S. presidential elections (Allcott and Gentzkow, 2017; Ward, 2018). Against this background, it is hardly surprising that there has been an increased interest in the development of methods, measures and computational tools that efficiently and effectively detect disinformation using machine learning and deep learning techniques. Among different approaches to fake news detection, language-based approaches have emerged as promising (for more details, see Section 2). Here the term 'language-based' is used in a broad sense to include a variety of approaches, such as those that employ traditional linguistic features, readability features, style-based features, discourse and rhetorical features or those that draw on word embedding techniques. The latter have proven to be particularly successful in detecting fake news. Despite their success, however, their detection is based on latent features that are not human interpretable and thus cannot explain why a piece of news was detected as fake news. As recently pointed out by Shu et al. (2019), white-box (explainable) approaches to fake news detection are desirable, since model-derived explanations can (1) provide valuable insights originally hidden to different stakeholders, such as policy makers, professional journalists and citizens and (2) can contribute to further improvement of fake news detection systems. This paper seeks to respond to recent calls for more explainable (white-box) approaches to fake news detection by performing a series

---

Proceedings of the 3rd International Workshop on Rumours and Deception in Social Media (RDSM), pages 14–31
Barcelona, Spain (Online), December 12, 2020.

of experiments where bi-directional recurrent neural network classifiers were trained on interpretable features derived from multi-disciplinary integrated approaches to language. The data come from two benchmark datasets and fake news detection is formulated as a binary classification and as a multiclass classification tasks correspondingly. The results of our experiments are promising, as our classification models achieve similar performance as the best-performing black box models reported in the literature. In a second step we report on ablation experiments geared towards assessing the relative importance of the human-interpretable features in distinguishing fake news from real news. The remainder of the paper is organized as follows: After a concise overview of related work in Section 2, Section 3 introduces the two data sets, Section 4 describes our approach to automated text analysis and six groups of language features used in the paper, Section 5 describes the model architecture, the training procedure and the method used to assess the relative feature importance. Sections 6 presents and discusses the main results and concluding remarks follow in Section 7.

2   Related Work

Here we provide a concise overview of recent approaches geared towards fake news detection that employ machine learning and deep learning techniques and we focus in particular on language-based approaches that are most pertinent to the purposes of this paper (for a more systematic and comprehensive overviews, see recent reviews and surveys by Shu et al. (2018), Oshikawa et al. (2020), Zhang and Ghorbani (2020) and Zhou and Zafarani (2020). Fake news detection is most often formulated as a binary classification task. However, categorizing all the news into two classes (fake vs real) is not the only conceivable way, since there are cases where the news is partially real and partially fake. A common practice is to add more classes distinguishing between several degrees of truthfulness and thus formulating fake news detection as a multi-class classification task. As will become evident later in this paper, we apply our approach to both scenarios. Three approaches to fake news detection frequently described in the literature are: (1) knowledge-based fake news detection (commonly using techniques from information retrieval to determine the veracity/truthfulness of news), (2) language-based fake news detection (drawing on traditional linguistic, style-related, readability or rhetorical features or using word embedding methods to distinguish between fake and real news) and (3) propagation-based fake news detection (typically using network analyses to determine the credibility of news sources at various stages, being created, published online and their spread via social media). Compared to knowledge-based and propagation-based approaches, language-based approaches are advantageous for several reasons, including: (1) they enable near real-time feedback (proactive rather than retroactive), i.e. they are not restricted to being applied only *a posteriori* (Potthast et al., 2017) and (2) they are scalable. A guiding assumption of language-based approaches is that there are statistical regularities inherent in natural languages and distributional patterns of language use indicative of fake news that are not consciously accessible to fake news creators. Space limitations prevent us from going into further details (but see reviews and survey cited above). In what follows, we will zoom in on previous studies on fake news detection conducted on the bases of the publicly available benchmark datasets used in the corpus study: the ISOT dataset, an 'entire article' dataset comprising 20k+ real and fake news texts (Ahmed et al., 2018), and the LIAR dataset, a 'claims dataset' comprising 12k+ real-world short statements collected from a variety of online sources (Wang, 2017) (see section 3 for details). Upon introduction of the ISOT dataset, (Ahmed et al., 2018) report on the results of experiments using n-gram features with two different features extraction techniques - Term Frequency (TF) and Term Frequency-Inverted Document Frequency (TF-IDF) - and six different machine learning techniques - Stochastic Gradient Descent, Support Vector Machines, Linear Support Vector Machines (LSVM), K-Nearest Neighbour and Decision Trees. Their best-performing model reached a classification accuracy of 92% using TF-IDF for feature extraction and an LSVM classifier, showing that real and fake news can be discriminated with high accuracy on the basis of the use of multiword sequences. However, subsequent studies have demonstrated that classification accuracy on this dataset can be pushed even higher - beyond the 99% accuracy mark - through the employment of deep neural networks trained on word embedding vectors: (Kula et al., 2020) reported classification accuracy between 95.04% and 99.86% using an LSTM neural network trained on different word embeddings (glove, news, Twitter, crawl) implemented in the Flair NLP framework (Akbik et al., 2019). Goldani et al., (2020) achieved a classification accuracy of 99.8% using a non-static capsule network and 'glove.6B.300d' word embeddings (Pennington et al., 2014). While the dataset ISOT involves a binary classification (fake vs. real), the LIAR dataset presents a six-way multiclass classification problem, where individual claims statement was evaluated for its truthfulness and received a much more fine-grained veracity label. In the experiments presented upon publication of the LIAR dataset, (Wang, 2017) provided several benchmarks based on several shallow learning classifiers (e.g. logistic regression and support vector machines) trained on n-gram features and deep learning classifiers (bi-directional long short-term memory and convolutional neural networks architectures) using pre-trained 300-dimensional word2vec embeddings from Google News (Mikolov et al., 2013). The latter reached a classification accuracy task of up to 27%. Incorporating available meta-data about the subject, speaker and context raised classification accuracy to 27.4%. Subsequent studies have shown that the classification accuracy on the LIAR set can be further increased to just over 45% by more complex hybrid models that integrate the linguistic information with speaker profiles into an attention based LSTM model (Long, 2017), by supplementing the data with verdict reports written by annotators (Karimi et al., 2018) or by replacing the credibility history in LIAR with a larger credibility source (Kirilin and Strube, 2018). Importantly, however, all state-of-the-art models designed to detect the veracity of a news article or claim exploit the information contained in high-dimensional word embeddings that are uninterpretable to humans, thereby severely limiting our ability to understand 'why' a given claim or news article was predicted to be fake or real.

3 Data

The experiments were conducted on two recently released datasets for fake news detection, the ISOT dataset compiled by the Information Security and Object Technology research lab (Ahmed et al., 2018) and the LIAR dataset introduced in (Wang, 2017). The datasets were selected based on their complementary attributes in terms of text types (full articles with average length of about 400 words vs. short statements with an average length of just under 20 words) and the granularity of the veracity labels (binary labels based on source selection and six-way classification based on ratings by politifac.com editors). Both datasets are sufficiently large for training deep models. The ISOT dataset consists of 40,000+ real and fake news articles collected from real-world sources between 2016 and 2017. The real (truthful) news articles were obtained by crawling articles from Reuters.com. The fake news articles were collected from unreliable websites that were flagged by politifact.com, a fact-checking organization in the USA, and Wikipedia. The ISOT dataset contains articles on a variety of topics with a focus on political and world news topics (see Table 2). For each article the following information is provided: article title, text, type (topic) and publication date. Close inspection of the dataset revealed that all and only instances of real news were introduced by the words "WASHINGTON (Reuters)", indicating the place and name of the news agency that has provided the news article. To prevent our models from capitalizing on this information, all instances of this string were deleted. We also checked for and removed all duplicates in the dataset (N = 6251). Table 2 presents the distribution of articles across news types (real/fake) and topics before and after deduplication (original/cleaned). The dataset was split in training, development, and testing sets using a 80/10/10 split. The LIAR dataset is a recent benchmark dataset for fake news detection that in includes 12,836 real-world short statements collected from a variety of online sources - including Facebook posts, tweets, news releases, TV/radio interviews, campaign speeches, TV ads and debates - on a range of topics - including economy, healthcare, taxes, federal-budget, education, jobs, state budget, candidates-biography, elections, and immigration. Each statement was labeled by an editor from politifact.com on a six-level ordinal scale of truthfulness ranging from "True", for completely accurate statements, to "Pants on Fire" (from the taunt "Liar, liar, pants on fire") for false and ludicrous claims. The distribution of the six labels is relatively well-balanced: with the exception of 1,050 instances of the 'pants-fire' category, the instances for all other labels range from 2,063 to 2,638. The LIAR set further includes a rich set of meta-data for each speaker including party affiliation, current job and home state. The statements in the dataset are also fairly balanced across the two major political parties of the US - democrats and republicans - and also contain a significant amount of posts from online social media. The dataset is distributed into training, validation and testing sets in a 80/10/10 manner.

4   Automated Text Analysis

The raw texts from the two datasets were automatically analyzed using CoCoGen, a computational tool that implements a sliding window technique to calculate within-text distributions of feature scores (see recently published papers that use this tool, (Ströbel et al., 2018; Kerz et al., 2020b; Kerz et al., 2020a). In contrast to the standard approach implemented in other tools for automated text analysis that rely on aggregate scores representing the average value of a feature in a text, the sliding-window approach generates a series of measurements representing the 'local' distributions of scores. A sliding window can be conceived of as a window of size ws, which is defined by the number of sentences it contains. The window is moved across a text sentence-by-sentence, computing one value per window for a given feature. The series of measurements faithfully captures a typically non-uniform distribution of features within a text and is referred here to as a 'contour'.[1] To compute the value of a given feature in a given window m $(w(m))$, a measurement function is called for each sentence in the window and returns a fraction (wn_m/wd_m). CoCoGen uses the Stanford CoreNLP suite (Manning et al., 2014) for performing tokenization, sentence splitting, part-of-speech tagging, lemmatization and syntactic parsing (Probabilistic Context Free Grammar Parser (Klein and Manning, 2003)). In its current version, CoCoGen supports a total 154 of features that fall into six categories: (1) features of syntactic complexity (N=19), (2) features of lexical density, sophistication and variation (N=12), (3) information-theoretic features (N=3), (4) register-based n-gram frequency features (N=25), (5) LIWC-style (Linguistic Inquiry and Word Count) features (N=61) and (6) Word-Prevalence measures (N=36). A brief overview of the features and their short descriptions are provided in Table 4 in the Appendix. The inclusion of these features[2] is motivated by contemporary language and cognitive sciences characterized by an integrated, multi-method, and transdisciplinary approach needed to advance our understanding of the human processing and learning mechanisms (Christiansen and Chater, 2017). The first three sets of features are derived from the literature on language development showing that, in the course of their lifespan, humans learn to produce and understand complex syntactic structures, more sophisticated and diverse vocabulary and informationally denser language (see, e.g., Berman, 2007; Lu, 2010, 2012; Hartshorne and Germine, 2015; Ehret and Szmrecsanyi, 2019). The fourth set of features is derived from research on language adaptation (Chang et al., 2012) and research that looks at language from the perspective of complex adaptive systems (Beckner et al., 2009; Christiansen and Chater, 2016) indicating that, based on accumulated language knowledge emerging from lifelong exposure to various types of language inputs, humans learn to adapt their language to meet the functional requirements of different communicative contexts. The features in set five are based on insights from many years of research conducted by Pennebaker and colleagues (Pennebaker et al., 2003; Tausczik and Pennebaker, 2010), showing that the words people use in their everyday life provide important psychological cues to their thought processes, emotional states, intentions, and motivations. And finally, the inclusion of features in group six is motivated by recent efforts to estimate of of how well words are known in the population through crowdsourcing and corpus-based techniques. An accumulating body of evidence shows that such word prevalence measures are good predictors of human perfomance on various language tasks (Brysbaert et al., 2019; Johns et al., 2020)

5   Classification Models

For the classification, we used Bi-directional Recurrent Neural Network (BRNN) classifiers with Gated Recurrent Unit (GRU) cells (Cho et al., 2014). BRNNs have been shown to outperform unidirectional RNNs in application areas ranging from acoustic modeling (Sak et al., 2014) to machine translation (Bahdanau et al., 2014). Bi-directional neural network models have also been employed in previous

---

[1]In general, for a text comprising n sentences, there are $w = n - ws + 1$ windows. Given the constraint that there has to be at least one window, a text has to comprise at least as many sentences at the ws is wide $n \geq w$.

[2]CoCoGen was designed with extensibility in mind, so that additional features can easily be implemented. It uses an abstract measure class for the implementation of additional features.

studies on the two datasets investigated here (Wang, 2017; Kula et al., 2020), making them well suited for purposes of comparison and, more specifically, for examining whether and to what extent a classifier trained on human-interpretable features can approximate the performance of a state-of-the-art classifier trained on word embeddings. Since the two datasets differ in terms of the availability of meta-data (ISOT: no meata-data, LIAR: rich information on subject, speaker and context) and with respect to the granularity at which truthfullness was assessed (ISOT: binary, LIAR: 6-way multiclass), the BRNN classifiers were adapted so as to take these differences into account.Figure 1 shows the architecture of models used in the present paper. $X = (x_1, x_2, \ldots, x_n)$ is the output from CoCoGen, which is a sequence of 154-dimensional vectors. To integrate the context information, the words in the context description were mapped to 300-dimensional word embedding vectors using the dependency based word-embedding implemented in spaCy (Honnibal and Montani, 2017), represented by $C = (c_1, c_2, \ldots, c_n)$. Instead of one-hot encoding, we use word embeddings and BRNN to encode the context meta information here, as otherwise the feature vector for context information would result in 5075-dimensional sparse one-hot vectors. $J = (j_1, j_2, \ldots, j_n)$ is a sequence of word embeddings for the job title of the speaker of a given text, following the same reasoning as above. $S = (s_1, s_2, \ldots, j_n)$ and $P = (p_1, s_2, \ldots, p_n)$ are 70 and 25 dimensional on-hot vector for state information and party affiliation of the speaker. The structure of the classifier for ISOT dataset is shown in 1 on the left hand side in Figure 1. The lower part encircled by the dashed red line represents the recurrent network, where the CoCoGen output for a given text is fed into a 2-layer BRNN consisting of GRU cells with 200 hidden units in each layer. h_{10}, h_{20} represent the initial hidden states of the first and second layer of the BRNN respectively in the forward direction and h'_{10}, h'_{20} represent the initial hidden states of the first and second layer of BRNN respectively in the backward direction. h_{2n} and h'_{2n} represent the last hidden states of the second layer of the BRNN in the forward and backward direction respectively. These layers are concatenated and passed through a feed-forward neural network, encircled by the blue dashed line in Figure 1. This network consists of three linear layers, whose output dimensions are 200, 100 and 2. Between layers 1 and 2 as well as between layers 2 and 3 we inserted a Batch Normalization (BN) layer, a Parametric ReLU (PReLU) activation function layer and a Dropout layer with a dropout rate of 0.5. A softmax layer is applied before the final output $\hat{y}$. For the LIAR dataset, we built three BRNN models: (1) a model using only the CoCoGen output (X), (2) a model using CoCoGen output and the context information (X + C) and (3) a model using CoCoGen output, the context information and the speaker profile, which comprises information about the job, the state and the party of the of a speaker (X + C + J + S + P). The structure of CoCoGen-only model is identical to model built for the ISOT dataset, with the exception that the output layer has a size of 6 instead of 2. In the CoCoGen + Context model shown in sub-figure 2 in Figure 1, the sequence vector $X = (x_1, x_2, \ldots, x_n)$ represents the CoCoGen output as described for the ISOT model above. BRNN blocks in sub-figure 2 has a same structure as the lower part of sub-figure 1, which is a 2-layer bidirectional RNN, whose output is a concatenation of the last hidden state of uppermost layer in forward and backward direction respectively. The BRNN on the left size in sub-figure 2 has a hidden state size of 200, while the BRNN on the right side has one of 10. The Feed-forward 1 block is identical to the Feed-forward part shown in sub-figure 1. Sub-figure 3 shows the structure of model making use of CoCoGen features + context + speaker profiles. S and P are one-hot encoded vectors described as above. They are squeezed to 10-dimensional vectors through a feed-forward neural network, Feed-forward 2, whose structure is shown in the lower right part of Figure 1. Feed-forward 2 consists of two linear layers, the output of which are 20 for Linear 1 and 10 for Linear 2 respectively. The BRNN for CoCoGen output and context are identical to the corresponding BRNN blocks mentioned above. The BRNN for job title information encoding has the same structure and hidden state size as BRNN for context. All output from BRNN blocks and Feed-forward 2 blocks are concatenated and fed into Feed-forward 1 block, whose structure is shown in the upper part of sub-figure **1** with the exception that linear layers have output size of 210, 105 and 5 respectively. Since the labels of the LIAR dataset are ordinal in nature, i.e. pants-fire < false < barely-true < half-true < mostly-true < true, the classification of instance in liar dataset can be treated as an ordinal classification problem. To adapt the neural network classifier to the ordinal classification task, we followed the NNRank approach described in (Cheng et al., 2008), which

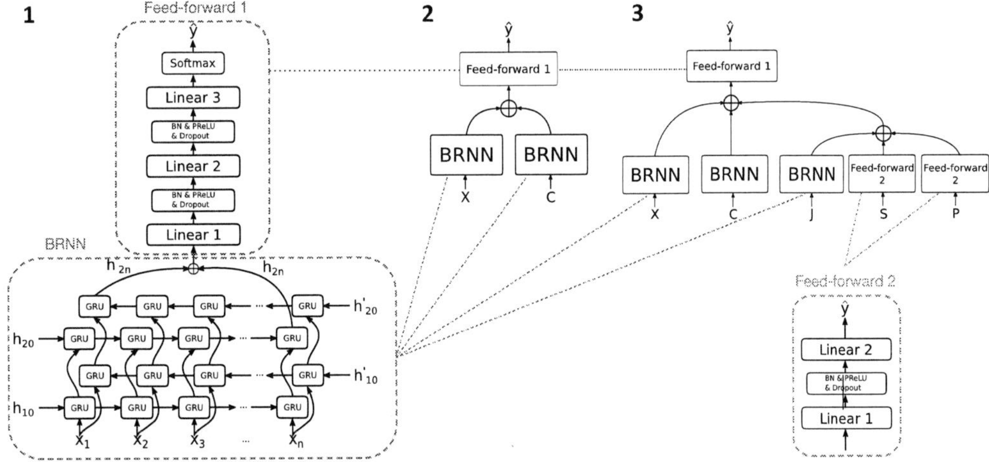

Figure 1: Structure of the BRNN classifiers built for the ISOT and LIAR datasets: The structure in 1 represents the model architecture used for ISOT and LIAR that makes use of textual information only (all CoCoGen features). The structures in 2 and 3 represent the model extensions that incorporate contextual meta-data (C) and speaker profiles (J = job title, P = party affiliation, S = speaker).

is a generalization of ordinal perception learning in neural networks(Crammer and Singer, 2002) and outperforms a neural network classifier on several benchmark datasets. Instead of one-hot encoding of class labels and using softmax as the output layer of a neural network, in NNRank, a class label for class k is encoded as $(y_1, y_2, \ldots, y_i, \ldots, y_{C-1})$, in which $y_i = 1$ for $i \leq k$ and $y_i = 0$ otherwise, where C is the number of classes. For the output layer, a sigmoid function was used. For prediction, the output of the neural network $(o_1, o_2, \ldots, o_{C-1})$ is scanned from left to right. It stops after encountering o_i, which is the first element of the output vector that is smaller than a threshold T (e.g. 0.5), or when there is no element left to be scanned. The predicted class of the output vector is the index k of the last element, whose value is greater than or equal to T. Finally, for the purpose of comparison, we also recreated the convolutional neural network (CNN) model described in (Wang, 2017). This CNN model consists of filters of size 2, 3 and 4. Each size has 128 filters with a max-pooling operation being performed on each output filter. The result of the max-pooling was fed into a feed-forward neural network for the classification. As an additional baseline, we further built structurally equivalent BRNN classifiers based on sentence embeddings from Sentence-BERT (SBERT) (Reimers and Gurevych, 2019).[3]

All models are implemented using PyTorch (Pytorch, 2019). For the BRNNs and the CNN that don't use the ordinal information cross entropy loss was used as a loss function:

$$\mathcal{L}(\hat{Y}, c) = -\sum_{i=1}^{C} p(y_i) \log(p(\hat{y}_i))$$

where c is the true class label of the current observation, C is the number of classes, $(p(y_1), \ldots, p(y_C))$ is a one-hot vector with

$$p(y_i) = \begin{cases} 1 & i = c \\ 0 & \text{otherwise} \end{cases}$$

and $\hat{Y} = (p(\hat{y}_1), p(\hat{y}_2), \ldots, p(\hat{y}_C))$ is the output vector of the softmax layer, which can be viewed as the predicted probabilities of the observed instance falling into to each of the classes. For training BRNNs

---

[3]SBERT is a finetuned BERT network using siamese and triplet network structures that. It has been shown to outperform other state-of-the-art sentence embeddings methods on common semantic textual similarity and transfer learning tasks (Reimers and Gurevych, 2019).

using ordinal information binary cross entropy was used:

$$\mathcal{L}(\hat{Y}, c) = -\frac{1}{C} \sum_{i=1}^{C} (y_i \log(\hat{y}) + (1 - y_i) \log(1 - \hat{y}))$$

in which $c = (y_1, y_2, \ldots, y_N)$, $C = 14$ is number of responses and $\hat{Y} = (\hat{y}_1, \hat{y}_2, \ldots, \hat{y}_N)$ is the output vector of the sigmoid layer rounded to closest integer. We tuned all hyperparameters on the validation set using a grid search over sets of optimizers $S = \{\text{Adamax}, \text{Adagrad}, \text{RMSprop}\}$, learning rates $L = \{0.01, 0.001, 0.0001\}$ and normalization methods $N = \{\text{Standardization}, \text{Min-max}\}$. The optimal hyperparameter combinations are provided in Table 5 in the Appendix.

To determine the relative importance of the language features groups, we conducted feature ablation experiments. Classical forward or backward sequential selection algorithms that proceed by sequentially adding or discarding features require a quadratic number of model training and evaluation in order to obtain a feature ranking (Langley, 1994). In the context of neural network models, training a quadratic number of models can become prohibitive. To alleviate this problem, we used an adapted version of the iterative sensitivity-based pruning algorithm proposed by (Díaz-Villanueva et al., 2010). This algorithm ranks the features based on a 'sensitivity measure' (Moody, 1994; Utans and Moody, 1991) and removes the least relevant variables one at a time. The classifier is then retrained on the resulting subset and a new ranking is calculated over the remaining features. This process is repeated until all features are removed. In this fashion, rather than training $\frac{n(n+1)}{2}$ models required for sequential algorithms, the number of models trained is reduced to $\frac{n}{m}$, where m is the number of features or feature groups that can be removed at each step. We report the results obtained after the removal of a single feature group at each step. At step t, a neural network model M_t is trained on the training set. The training set at step t consists of instances with feature groups $F_t = \{f_1, f_2, \ldots, f_{D_t}\}$ where $f_1, \ldots f_{D_t}$ are the remaining feature groups at the current step, whose importance rank is to be determined. We define X_t as the test set with feature set F_t and X_t^i as the same dataset as X_t except we set the ith feature f_i of each instance within the dataset to its average. Furthermore, we define $g(X)$ as the classification accuracy of $M_{t,n}$ for a dataset X. The sensitivity of a feature group f_i at step t is obtained from:

$$S_{i,t} = g(X_t) - g(X_t^i)$$

The most important feature group at step t can be found by:

$$f_{\hat{i}} : \hat{i} =_{i:f_i \in F_t} (S_{i,t})$$

Then we set the rank for feature $f_{\hat{i}}$:

$$Rank_{\hat{i}} = t$$

In the end, feature $f_{\hat{i}}$ is dropped from F_t and the corresponding columns in training and test dataset are also dropped simultaneously:

$$F_{t+1} = F_t - \{f_{\hat{i}}\}$$

This procedure is repeated, until $|F_{t'}| = 1$.

6 Results

The performance metrics of the classification models for both datasets (global accuracy, precision and recall) are presented in Table 1, along with comparisons with the results of previous studies (a extended version of the table with performance data of additional models is provided in the Appendix). The results of our BRNN classifiers trained on interpretable features are highly competitive with those obtained from state-of-the-art RNN, CNN and capsule networks that exploit word embeddings to represent textual contents. In fact, in both datasets, our classifiers match the performance of the best-performing models within half a percent: For ISOT, the CAPSULE-glove (Goldani et al., 2020) and LSTM-glove (Kula et al., 2020) both achieve an accuracy of 99.8%, while BRNN CoCoGen achieves 99.3%. Moreover, the BRNN CoCoGen model outperformed the LSTMs presented in Kula et al. (2020) that utilize three other word embeddings implemented in the Flair library (news, Twitter, crawl) by up to 4.3% and improved on the performance on the n-gram-based LSVM model by 7.3%. For the LIAR data set, the difference in classification accuracy between BRNN CoCoGen and the CNN utilizing 300-dimensional word embeddings trained on Google News presented in Wang (2017) amounts to 0.2%, when meta-data on context and speaker profiles is taken into account. Excluding all meta-data, the BRNN CoCoGen (ordered) model reached an accuracy of 27.7%, which is even slightly higher than the performance of the Bi-LSTM 300-dim word2vec embeddings (Google news) model. Our CNN CoCoGen model achieved a classification accuracy of 25.6%, which is 1.4% below the performance of the corresponding CNN model presented in Wang (2017) , CNN 300-dim word2vec embeddings (Google News). Interestingly, however, this model suffered from a substantial drop in accuracy to 24.8%, once it was infused with contextual meta-data. In contrast, all BRNN CoCoGen models invariably benefited from the addition of any type of meta-data. While performance with the CAPSULE-glove networks presented in Goldani et al. (2020) is limited by their selective integration of meta-data, it is worth noting that the CNN CoCoGen model outperformed all their models without recourse to meta-data. Taken together these results present strong evidence that successful detection of fake news can be achieved without sacrificing transparency. It is also worth pointing out that approaching the fake news detection task as an ordinal classification problem had considerable effects on a classifiers performance. Specifically, we observed (1) that classification accuracy slightly increased by 0.6% relative to a unordered classification approach and (2) that classification behavior shifted from a bias towards recall to a bias towards precision. Furthermore, comparison of the confusion matrices of our classifiers revealed that changing to the ordinal classification approach had positive effects on the distribution of errors: The ordinal classification problem is monotonic, meaning that the further a misclassification is from the main diagonal of a confusion matrix, the more severe it is. The confusion matrix of the best-performing BRNN CoCoGen model shows that for five out of the six classes (pants-fire, false, half-true, mostly-true, true) the most frequent prediction was the true class and the number of misclassifications decreases with increasing distance to the true class. In contrast, in the case of the unordered classifiers, we observed that the extreme categories ('pants-fire' and 'true') were avoided and predictions to the intermediate categories were preferred, especially in classifiers without meta-data information (confusion matrices for all models are provided in the Appendix). To the best of the authors knowledge, current models on multi-class fake news detection do not concern with the order of labels (Oshikawa et al., 2018). Our results indicate that future work can benefit from taking an ordinal classification approach. The results of our feature ablation experiments revealed a similar rank order in feature importance in both datasets (detailed results can be found in Table 14 in the Appendix): In each case, classification performance was mainly driven by features from the groups Lexical, LIWC, Syntactic and register-based n-grams, and to a lesser extent by information theoretic and word-prevalence-based features. Specifically, Table 14 indicates that - in the casse of the ISOT dataset - dropping the features from the LIWC group results in the largest decrease in classification accuracy of 5.1% on the validation set, resulting in a drop in accuracy on the test set to 93.8%. Re-training the model without the LIWC features yields the new baseline of 99.1%, indicating that the remaining features contained enough information to allow the retrained model to compensate for the loss of the LIWC information. After the elimination of the next two most-important feature groups, the syntactic and lexical groups, the retrained model at iteration 3 is still able to achieve an accuracy on the validation set of 97.9%. However, after the drop of the n-gram feature group, classification accuracy on the drops to 76.3% (validation) and 76.2% (test), indicating that the lost information from the four top-feature groups cannot be compensated for by information from the remaining feature groups, i.e. information theoretic and word-prevalence-based features. In the case of the the LIAR dataset, the relative influence of the six feature groups is more even and the predictive power of the model (27.2% accuracy on the test set) appears to stem from exploiting information from all six feature groups. For a closer examination of how individual features within each feature-group distinguished between real and fake news, we derived standard scores by performing z-standardization on all indicators and determined the difference between mean standard scores of real and fake news ($DeltaScore_{index\ i} = Score_{index\ i,\ fake\ news} - Score_{index\ i,\ real\ news}$) (a complete table with the DeltaScores for the top-20 features for both datasets is provided in the Appendix). Inspection

Dataset	Model	Validation set			Test set		
		Accuracy	Precision	Recall	Accuracy	Precison	Recall
ISOT	LSVM unigram 50k[1]	–	–	–	0.920	–	–
	LSTM-glove[2]	–	–	–	**0.998**	–	–
	CAPSULE-glove[3]	–	–	–	**0.998**	–	–
	BRNN SBERT	0.998	0.998	0.998	0.997	0.997	0.997
	BRNN CoCoGen	0.994	0.994	0.994	**0.993**	0.993	0.993
LIAR	Bi-LSTM 300-dim word2vec[4] embeddings (Google News)	0.223	–	–	0.233	–	–
	CNN 300-dim word2vec[4] embeddings (Google News) + context + speaker profile	0.247	–	–	**0.274**	–	–
	CAPSULE-glove + Party[3]	0.261	–	–	0.240	–	–
	CAPSULE-glove + State[3]	0.240	–	–	0.243	–	–
	CAPSULE-glove + Job[3]	0.254	–	–	0.251	–	–
	BRNN SBERT (ordered)	0.292	0.272	0.327	0.270	0.296	0.249
	BRNN CoCoGen (ordered)	0.251	0.281	0.218	0.237	0.217	0.207
	BRNN CoCoGen (ordered) + context	0.264	0.280	0.241	0.253	0.281	0.238
	BRNN CoCoGen (ordered) + context + speaker profile	0.284	0.305	0.263	**0.272**	0.304	0.258

Table 1: Evaluation results on the ISOT and LIAR datasets on the validation and test sets. [1] = Ahmed et al., 2018; [2] = Kula et al., 2020; [3] = Goldani et al., 2020; [4] = Wang, 2017

of the Delta Scores revealed some interesting patterns. For example, real news articles and claims are characterized by (1) relatively higher lexical diversity (as measured by type-token ratio features), (2) stronger reliance of multiword sequences from the news and academic register (measured by register-based n-gram frequency measures), (3) greater phrasal syntactic complexity (as measured, e.g., by the number of complex nominals per clause) and (4) more frequent use or word from particular domains, such as work, money, power or word classes, such as preposition and quantifiers. In contrast, fake news are characterized by (1) greater syntactic complexity (as measured by, e.g. by the number of clauses per sentence), (2) frequent use of multiword sequences form the domain of fiction, (3) higher lexical sophistication scores (as measured in terms of relatively infrequent words) and (4) a strong reliance on personal pronouns, adverbs and emotion words. While limitations of space preclude an in-depth discussion, these results demonstrate that the use of interpretable features can provide new insights and knowledge about the characteristics of fake news and explain "why" a piece of news was detected as fake news see (Shu et al., 2019) for a discussion of explainable fake news detection.

7   Conclusion and Future Work

In recent years, there is a growing recognition of the need to move away from black-box models towards white-box models for solving practical problems, in particular in the context of critical industries, including healthcare, criminal justice, and news (Rudin, 2019). This is due to the fact that human experts in a given application domain need both accurate but also understandable models (Loyola-Gonzalez, 2019). In this paper, we have made a contribution to this development in the domain of fake news detection. We have demonstrated that models trained on human interpretable features in combination with deep learning classifiers can compete with black box models based on word embeddings. In the future we intend to extend this work in two directions: First, we plan to apply our approach to fake news detection in German whose research still lags far behind that available for English. Second, we also plan to apply our approach to the detection of rumours and conspiracy theories to tackle and combat the ongoing Covid-19 infodemic.

References

Hadeer Ahmed, Issa Traore, and Sherif Saad. 2018. Detecting opinion spams and fake news using text classification. *Security and Privacy*, 1(1):e9.

Alan Akbik, Tanja Bergmann, Duncan Blythe, Kashif Rasul, Stefan Schweter, and Roland Vollgraf. 2019. Flair: An easy-to-use framework for state-of-the-art nlp. In *Proceedings of the 2019 Conference of the North American Chapter of the Association for Computational Linguistics (Demonstrations)*, pages 54–59.

Hunt Allcott and Matthew Gentzkow. 2017. Social media and fake news in the 2016 election. *Journal of economic perspectives*, 31(2):211–36.

Dzmitry Bahdanau, Kyunghyun Cho, and Yoshua Bengio. 2014. Neural machine translation by jointly learning to align and translate. *arXiv preprint arXiv:1409.0473*.

Clay Beckner, Richard Blythe, Joan Bybee, Morten H Christiansen, William Croft, Nick C Ellis, John Holland, Jinyun Ke, Diane Larsen-Freeman, et al. 2009. Language is a complex adaptive system: Position paper. *Language learning*, 59:1–26.

Ruth A Berman. 2007. Developing linguistic knowledge and language use across adolescence.

Marc Brysbaert, Paweł Mandera, Samantha F McCormick, and Emmanuel Keuleers. 2019. Word prevalence norms for 62,000 english lemmas. *Behavior research methods*, 51(2):467–479.

Franklin Chang, Marius Janciauskas, and Hartmut Fitz. 2012. Language adaptation and learning: Getting explicit about implicit learning. *Language and Linguistics Compass*, 6(5):259–278.

Jianlin Cheng, Zheng Wang, and Gianluca Pollastri. 2008. A neural network approach to ordinal regression. In *2008 IEEE International Joint Conference on Neural Networks (IEEE World Congress on Computational Intelligence)*, pages 1279–1284. IEEE.

KyungHyun Cho, Bart van Merrienboer, Dzmitry Bahdanau, and Yoshua Bengio. 2014. On the properties of neural machine translation: Encoder-decoder approaches. *CoRR*, abs/1409.1259.

Morten H Christiansen and Nick Chater. 2016. *Creating language: Integrating evolution, acquisition, and processing*. MIT Press.

Morten H Christiansen and Nick Chater. 2017. Towards an integrated science of language. *Nature Human Behaviour*, 1(8):1–3.

Koby Crammer and Yoram Singer. 2002. Pranking with ranking. In *Advances in neural information processing systems*, pages 641–647.

Wladimiro Díaz-Villanueva, Francesc J Ferri, and Vicente Cerverón. 2010. Learning improved feature rankings through decremental input pruning for support vector based drug activity prediction. In *International Conference on Industrial, Engineering and Other Applications of Applied Intelligent Systems*, pages 653–661. Springer.

Katharina Ehret and Benedikt Szmrecsanyi. 2019. Compressing learner language: An information-theoretic measure of complexity in sla production data. *Second Language Research*, 35(1):23–45.

Mohammad Hadi Goldani, Saeedeh Momtazi, and Reza Safabakhsh. 2020. Detecting fake news with capsule neural networks. *arXiv preprint arXiv:2002.01030*.

Joshua K Hartshorne and Laura T Germine. 2015. When does cognitive functioning peak? the asynchronous rise and fall of different cognitive abilities across the life span. *Psychological science*, 26(4):433–443.

Matthew Honnibal and Ines Montani. 2017. spaCy 2: Natural language understanding with Bloom embeddings, convolutional neural networks and incremental parsing. To appear.

Brendan T Johns, Melody Dye, and Michael N Jones. 2020. Estimating the prevalence and diversity of words in written language. *Quarterly Journal of Experimental Psychology*, 73(6):841–855.

Hamid Karimi, Proteek Roy, Sari Saba-Sadiya, and Jiliang Tang. 2018. Multi-source multi-class fake news detection. In *Proceedings of the 27th International Conference on Computational Linguistics*, pages 1546–1557.

Elma Kerz, Fabio Pruneri, Daniel Wiechmann, Yu Qiao, and Marcus Ströbel. 2020a. Understanding the dynamics of second language writing through keystroke logging and complexity contours. In *Proceedings of The 12th Language Resources and Evaluation Conference (LREC2020))*, pages 182–188.

Elma Kerz, Yu Qiao, Daniel Wiechmann, and Marcus Ströbel. 2020b. Becoming linguistically mature: Modeling english and german children's writing development across school grades. In *Proceedings of the Fifteenth Workshop on Innovative Use of NLP for Building Educational Applications (BEA2020))*, pages 65–74.

Angelika Kirilin and Micheal Strube. 2018. Exploiting a speaker's credibility to detect fake news. In *Proceedings of Data Science, Journalism & Media workshop at KDD (DSJM'18)*.

Dan Klein and Christopher D Manning. 2003. Accurate unlexicalized parsing. In *Proceedings of the 41st annual meeting of the association for computational linguistics*, pages 423–430.

Sebastian Kula, Michał Choraś, Rafał Kozik, Paweł Ksieniewicz, and Michał Woźniak. 2020. Sentiment analysis for fake news detection by means of neural networks. In *International Conference on Computational Science*, pages 653–666. Springer.

Pat Langley. 1994. Selection of relevant features in machine learning. In *Proceedings of the AAAI Fall symposium on relevance*, pages 1–5.

Yunfei Long. 2017. Fake news detection through multi-perspective speaker profiles. Association for Computational Linguistics.

Octavio Loyola-Gonzalez. 2019. Black-box vs. white-box: Understanding their advantages and weaknesses from a practical point of view. *IEEE Access*, 7:154096–154113.

Xiaofei Lu. 2010. Automatic analysis of syntactic complexity in second language writing. *International Journal of Corpus Linguistics*, 15(4):474–496.

Xiaofei Lu. 2012. The relationship of lexical richness to the quality of ESL learners' oral narratives. *The Modern Language Journal*, 96(2):190–208.

Christopher D. Manning, Mihai Surdeanu, John Bauer, Jenny Finkel, Prismatic Inc, Steven J. Bethard, and David Mcclosky. 2014. The stanford corenlp natural language processing toolkit. In *In ACL, System Demonstrations*.

Tomas Mikolov, Kai Chen, Greg Corrado, and Jeffrey Dean. 2013. Efficient estimation of word representations in vector space. *arXiv preprint arXiv:1301.3781*.

John Moody. 1994. Prediction risk and architecture selection for neural networks. In *From statistics to neural networks*, pages 147–165. Springer.

Ray Oshikawa, Jing Qian, and William Yang Wang. 2018. A survey on natural language processing for fake news detection. *arXiv preprint arXiv:1811.00770*.

James W Pennebaker, Matthias R Mehl, and Kate G Niederhoffer. 2003. Psychological aspects of natural language use: Our words, our selves. *Annual review of psychology*, 54(1):547–577.

Jeffrey Pennington, Richard Socher, and Christopher D Manning. 2014. Glove: Global vectors for word representation. In *Proceedings of the 2014 conference on empirical methods in natural language processing (EMNLP)*, pages 1532–1543.

Martin Potthast, Johannes Kiesel, Kevin Reinartz, Janek Bevendorff, and Benno Stein. 2017. A stylometric inquiry into hyperpartisan and fake news. *arXiv preprint arXiv:1702.05638*.

Pytorch. 2019. Pytorch: Tensors and dynamic neural networks in python with strong gpu acceleration. `https://github.com/pytorch/pytorch`.

Nils Reimers and Iryna Gurevych. 2019. Sentence-BERT: Sentence embeddings using siamese BERT-networks. In *Proceedings of the 2019 Conference on Empirical Methods in Natural Language Processing*. Association for Computational Linguistics, 11.

Cynthia Rudin. 2019. Stop explaining black box machine learning models for high stakes decisions and use interpretable models instead. *Nature Machine Intelligence*, 1(5):206–215.

Hasim Sak, Andrew W Senior, and Françoise Beaufays. 2014. Long short-term memory recurrent neural network architectures for large scale acoustic modeling.

Kai Shu, Amy Sliva, Suhang Wang, Jiliang Tang, and Huan Liu. 2017. Fake news detection on social media: A data mining perspective. *ACM SIGKDD explorations newsletter*, 19(1):22–36.

Kai Shu, Limeng Cui, Suhang Wang, Dongwon Lee, and Huan Liu. 2019. defend: Explainable fake news detection. In *Proceedings of the 25th ACM SIGKDD International Conference on Knowledge Discovery & Data Mining*, pages 395–405.

Marcus Ströbel, Elma Kerz, Daniel Wiechmann, and Yu Qiao. 2018. Text genre classification based on linguistic complexity contours using a recurrent neural network. In *Proceedings of the Tenth International Workshop 'Modelling and Reasoning in Context' co-located with the 27th International Joint Conference on Artificial Intelligence (IJCAI 2018) and the 23rd European Conference on Artificial Intelligence*, pages 56–63.

Yla R Tausczik and James W Pennebaker. 2010. The psychological meaning of words: Liwc and computerized text analysis methods. *Journal of language and social psychology*, 29(1):24–54.

Joshua A Tucker, Andrew Guess, Pablo Barberá, Cristian Vaccari, Alexandra Siegel, Sergey Sanovich, Denis Stukal, and Brendan Nyhan. 2018. Social media, political polarization, and political disinformation: A review of the scientific literature. *Political polarization, and political disinformation: a review of the scientific literature (March 19, 2018)*.

Joachim Utans and John Moody. 1991. Selecting neural network architectures via the prediction risk: Application to corporate bond rating prediction. In *Proceedings First International Conference on Artificial Intelligence Applications on Wall Street*, pages 35–41. IEEE.

William Yang Wang. 2017. " liar, liar pants on fire": A new benchmark dataset for fake news detection. *arXiv preprint arXiv:1705.00648*.

Ken Ward. 2018. Social networks, the 2016 us presidential election, and kantian ethics: applying the categorical imperative to cambridge analytica's behavioral microtargeting. *Journal of media ethics*, 33(3):133–148.

Xichen Zhang and Ali A Ghorbani. 2020. An overview of online fake news: Characterization, detection, and discussion. *Information Processing & Management*, 57(2):102025.

Xinyi Zhou, Reza Zafarani, Kai Shu, and Huan Liu. 2019. Fake news: Fundamental theories, detection strategies and challenges. In *Proceedings of the twelfth ACM international conference on web search and data mining*, pages 836–837.

Appendix

Table 2: Composition of the ISOT dataset; sizes indicate the number of articles in a given category; 'cleaned' refers to the datasets after deduplication

News Type	Total size	Topic	Size
Real News	original: 21417; cleaned: 21192	World-News	original: 10145; cleaned: 9978
		Politics-News	original: 11272; cleaned: 11214
Fake-News	original: 23481: cleaned: 17455	Government-News	original: 1570; cleaned: 514
		Middle-east	original: 778; cleaned: 0
		US News	original: 783; cleaned: 783
		Left-News	original: 4459; cleaned: 683
		Politics	original: 6841; cleaned: 6425
		News	original: 9050; cleaned: 9050

Table 3: Composition of the LIAR dataset

Training set size	10,269
Validation set size	1,284
Testing set size	1,283
Avg. statement length (tokens)	17.9
Top-3 Speaker Aliations	
Democrats	4,150
Republicans	5,687
None (e.g., FB posts)	2,185

Table 4: Concise overview of the six feature groups

Feature group	Size	Subtypes	Example/Description
Syntactic complexity	18	Length of production unit Subordination Coordination Particular structures	e.g. mean length of clause e.g. clauses per sentences e.g. Coordinate phrases per clause e.g. Complex nominals per clause
Lexical richness	12	Lexical density Lexical diversity Lexical sophistication	e.g. ration contents words / all words e.g. type token ratio e.g. words on General Service List
Register-based n-gram frequency	25	Spoken ($n \in [1,5]$) Fiction ($n \in [1,5]$) Magazine ($n \in [1,5]$) News ($n \in [1,5]$) Academic ($n \in [1,5]$)	measures of frequencies of n-grams of order 1-5 from five language registers
Information theory	3	Kolmogorov$_{\text{Deflate}}$ Kolmogorov$_{\text{Deflate Syntactic}}$ Kolmogorov$_{\text{Deflate Morphological}}$	measures use Deflate algorithm and relate size of compressed file to size of original file
LIWC-style	60	2300 words from > 70 classes	classes include e.g. function, grammar perceptual, cognitive and biological processes, personal concerns, affect, social, basic drives, ...
Word-Prevalence	36	crowdsourcing-based corpus-based	measures capture information on word frequency, contextual diversity and semantic distinctiveness differentiated across language variety (US, UK) and gender (male, female)

Dataset	Model	Optimizer	learning rate	Normalization Method
ISOT	BRNN	Adamax	0.001	Standardization
LIAR	BRNN (ordered)	Adamax	0.001	Standardization
	BRNN (unordered)	RMSprop	0.001	Min-Max
	CNN	RMSprop	0.01	Standardization
	BRNN + context	Adamax	0.0001	Standardization
	BRNN + context + speaker profile (unordered)	RMSprop	0.0001	Standardization
	BRNN + context + speaker profile (ordered)	Adamax	0.001	Standardization

Table 5: Optimal combinations of optimizer, learning rate and normalization methods identified via grid search.

Dataset	Model	Validation set			Test set		
		Accuracy	Precision	Recall	Accuracy	Precison	Recall
ISOT	LSVM unigram 50k (Ahmed et al. 2018)	–	–	–	0.920	–	–
	LSTM-glove	–	–	–	**0.998**	–	–
	LSTM-news	–	–	–	0.950	–	–
	LSTM-twitter	–	–	–	0.980	–	–
	LSTM-crawl (Kula et al., 2020)	–	–	–	0.976	–	–
	CAPSULE-glove (Goldani et al., 2020)	–	–	–	**0.998**	–	–
	BRNN SBERT (ordered)	0.292	0.272	0.327	0.270	0.296	0.249
	BRNN CoCoGen	0.994	0.994	0.994	**0.993**	0.993	0.993
LIAR	Bi-LSTM 300-dim word2vec embeddings (Google News)	0.223	–	–	0.233	–	–
	CNN 300-dim word2vec embeddings (Google News)	0.260	–	–	0.270	–	–
	CNN 300-dim word2vec embeddings (Google News) + context	0.277	–	–	0.248	–	–
	CNN 300-dim word2vec embeddings (Google News) + context + speaker profile (Wang, 2017)	0.247	–	–	**0.274**	–	–
	CAPSULE-glove + Party	0.261	–	–	0.240	–	–
	CAPSULE-glove + State	0.240	–	–	0.243	–	–
	CAPSULE-glove + Job (Goldani et al., 2020)	0.254	–	–	0.251	–	–
	BRNN SBERT (ordered)	0.292	0.272	0.327	0.270	0.296	0.249
	BRNN CoCoGen (unordered)	0.269	0.186	0.227	0.244	0.172	0.207
	BRNN CoCoGen (ordered)	0.251	0.281	0.218	0.237	0.217	0.207
	CNN CoCoGen (unordered)	0.266	0.357	0.224	0.256	0.155	0.216
	BRNN CoCoGen (ordered) + context	0.264	0.280	0.241	0.253	0.281	0.238
	BRNN CoCoGen (unordered) context + speaker profile	0.288	0.233	0.253	0.266	0.217	0.231
	BRNN CoCoGen (ordered) + context + speaker profile	0.284	0.305	0.263	**0.272**	0.304	0.258

Table 6: Evaluation results on the ISOT and LIAR datasets on the validation and test sets. Models indexed as "CoCoGen" comprise textual features only. Models with "+" are hybrid models with textual and meta-data. The labels "ordered" and "unordered" indicate whether an ordinal and nominal classification method was applied.

Table 7: Add caption

Top-20 Measures Fake News

	LIAR		ISOT	
	Measure	Delta Score	Measure	Delta Score
1	Lexical Density	0.183	LIWC Adverb	0.894
2	LIWC Focus future	0.132	LIWC Ipron	0.733
3	LIWC Relig	0.117	ngram 2 fic	0.728
4	Lexical Div CNDW	0.115	LIWC You	0.681
5	Lexical Div TTR	0.115	LIWC Focuspresent	0.677
6	Lexical Soph BNC	0.114	Mor Kolmogorov	0.670
7	LIWC Verb	0.106	Syntactic ClausesPerSentence	0.664
8	LIWC Hear	0.099	Base Kolmogorov	0.652
9	MeanLengthWord	0.098	LIWC Certain	0.630
10	Base Kolmogorov	0.095	Syntactic Kolmogorov	0.629
11	LIWC Negate	0.091	ngram 3 fic	0.620
12	Lexical Soph ANC	0.091	LIWC See	0.561
13	LIWC Posemo	0.087	Syntactic DepClausesPerTUnit	0.502
14	Morphological Kolmogorov	0.087	LIWC Interrog	0.498
15	Syntactic Kolmogorov	0.084	LIWC I	0.438
16	Syntactic VerbPhrasesPerTUnit	0.081	LIWC They	0.431
17	MeanSyllablesPerWord	0.076	LIWC Shehe	0.420
18	Lexical Soph NGSL	0.075	LIWC Female	0.385
19	LIWC Risk	0.062	LIWC Swear	0.380
20	LIWC Focuspresent	0.059	ngram 1 fic	0.370

Top-20 Measures Real News

	LIAR		ISOT	
	Measure	Delta Score	Measure	Delta Score
1	LIWC Quant	-0.212	Syntactic ComplexNomPerClause	-0.940
2	LIWC Compare	-0.197	Syntactic MeanLengthClause	-0.929
3	LIWC Adj	-0.171	LIWC Prep	-0.763
4	ngram 2 news	-0.148	LIWC Hear	-0.747
5	ngram 2 acad	-0.147	LIWC Power	-0.739
6	ngram 2 mag	-0.145	LIWC Work	-0.734
7	LIWC Time	-0.133	LIWC Article	-0.723
8	WordPrevalence	-0.132	LIWC Focus past	-0.666
9	ngram 1 acad	-0.131	LIWC Space	-0.545
10	ngram 3 acad	-0.128	Syntactic CoordPhrasesPerClause	-0.509
11	ngram 3 mag	-0.123	NP PreModWords	-0.493
12	ngram 3 news	-0.122	Lexical Div TTR	-0.465
13	ngram 1 mag	-0.120	Lexical Div CNDW	-0.465
14	ngram 1 fic	-0.118	Lexical Div RTTR	-0.413
15	ngram 1 news	-0.118	Lexical Div CTTR	-0.403
16	LIWC Number	-0.116	LIWC Money	-0.314
17	ngram 1 spok	-0.111	Lexical Soph BNC	-0.309
18	LIWC Space	-0.110	ngram 5 news	-0.306
19	LIWC Prep	-0.106	LIWC Achieve	-0.303
20	ngram 2 spok	-0.105	Lexical Density	-0.302

	pants-fire	false	barely-true	half-true	mostly-true	true
pants-fire	1	23	24	31	11	2
false	2	49	55	84	53	6
barely-true	3	36	62	80	27	4
half-true	1	38	58	108	55	5
mostly-true	1	19	41	104	74	2
true	3	19	43	71	66	6

Table 8: confusion matrix of liar dataset BRNN model

	pants-fire	false	barely-true	half-true	mostly-true	true
pants-fire	0	43	2	31	16	0
false	0	107	4	65	73	0
barely-true	0	78	8	61	65	0
half-true	0	83	9	84	89	0
mostly-true	0	44	3	84	110	0
true	0	60	0	60	88	0

Table 9: confusion matrix of liar dataset BRNN model (non-ordinal)

	pants-fire	false	barely-true	half-true	mostly-true	true
pants-fire	0	42	1	29	20	0
false	0	114	3	57	75	0
barely-true	0	86	2	71	52	1
half-true	0	88	4	97	76	0
mostly-true	0	77	2	51	111	0
true	0	74	1	46	87	0

Table 10: confusion matrix of liar dataset CNN model

	pants-fire	false	barely-true	half-true	mostly-true	true
pants-fire	15	29	22	14	11	1
false	8	57	65	62	49	8
barely-true	8	38	62	66	33	5
half-true	5	35	72	99	50	4
mostly-true	3	22	50	86	77	3
true	3	18	51	68	58	10

Table 11: confusion matrix of liar dataset BRNN model meta context

	pants-fire	false	barely-true	half-true	mostly-true	true
pants-fire	20	27	18	14	12	1
false	8	54	64	74	44	5
barely-true	5	39	59	65	42	2
half-true	2	32	46	111	69	5
mostly-true	1	21	36	85	94	4
true	3	25	31	64	79	6

Table 12: confusion matrix of liar dataset BRNN model all meta

	pants-fire	false	barely-true	half-true	mostly-true	true
pants-fire	0	49	17	9	14	3
false	0	108	20	54	38	29
barely-true	0	73	27	55	42	15
half-true	0	76	23	83	61	22
mostly-true	0	43	15	70	88	25
true	0	57	12	45	63	31

Table 13: confusion matrix of liar dataset BRNN model all meta (non-ordinal)

Dataset	Feature Group	Accuracy base model (validation)	Accuracy after drop (validation)	Accuracy after drop (test)
ISOT	LIWC	0.993	0.942	0.938
	Syntactic	0.991	0.964	0.965
	Lexical	0.988	0.915	0.912
	N-grams	0.989	0.763	0.762
	Info theory	0.979	0.822	0.823
	Word-prevalence	0.933	0.482	0.475
LIAR	Lexical	0.255	0.217	0.209
	LIWC	0.252	0.204	0.192
	Syntactic	0.232	0.193	0.215
	N-grams	0.224	0.188	0.218
	Word-prevalence	0.210	0.190	0.205
	Info theory	0.209	0.193	0.208

Table 14: Results of the feature ablation experiments for the ISOT dataset (top) and the LIAR dataset (bottom).

Covid or not Covid? Topic Shift in Information Cascades on Twitter

Liana Ermakova
HCTI - EA 4249
Université de Bretagne Occidentale
Brest, France
liana.ermakova@univ-brest.fr

Diana Nurbakova
LIRIS UMR 5205 CNRS
INSA Lyon
University of Lyon
Villeurbanne, France
diana.nurbakova@insa-lyon.fr

Irina Ovchinnikova
Sechenov First Moscow State
Medical University
Moscow, Russia
ovchinnikova.ig@1msmu.ru

Abstract

Social media have become a valuable source of information. However, its power to shape public opinion can be dangerous, especially in the case of misinformation. The existing studies on misinformation detection hypothesise that the initial message is fake. In contrast, we focus on information distortion occurring in cascades as the initial message is quoted or receives a reply. We show a significant topic shift in information cascades on Twitter during the Covid-19 pandemic providing valuable insights for the automatic analysis of information distortion.

1 Introduction

Social media is a valuable resource for all sorts of information. However, its power to shape public opinion can provoke serious societal issues such as misinformation. Words or actions by a *Public Figure (PF)* generate 69% of misinformation in discussions with ordinary users, while PFs themselves are responsible for 20% of the messages containing distorted information (Brennen et al., 2020). PFs post tweets that are likely to be shared by their followers (Romero et al., 2011), thus generating information cascades. The periodic repetitions provoke the mutability of information diffusion in the political domain (Shin et al., 2018). Recent studies show a similarity of distorted information dissemination during the pandemic to the distribution of political misinformation (Pennycook et al., 2020; Pennycook and Rand, 2018). The repetitions of rumours about conspiracy theories associated with Covid-19 led to the mutability of information; nevertheless, many users ridiculed these theories while repeating the rumors (Ahmed et al., 2020). During the Covid-19 pandemic, users look for medical information in PF feeds and follow personal stories of infected people who share unverified information because complicated medical texts deter lay readers (Ribeiro et al., 2019). Mass medical information sharing generates cascades where the probability to distort initial information increases due to omissions and paraphrases. As medical discourse is sensitive to any changes in terminology and text structure made by incompetent people (Nye et al., 2018), the impact of medical misinformation on social behaviour means that there is a pressing need to understand how it circulates on social media. To reveal crucial issues about Covid-19 that are of importance for lay people we need to understand topic shifts occurring within information cascades about the pandemic. Such understanding allows us to discover a particular lack of medical information and demand for clear explanation of the most important public problems of the current pandemic. In this paper, we present a preliminary study on medical information distortion occurring in cascades on Twitter due to topic shift. Several studies have focused on misinformation during the Covid-19 pandemic (Pennycook et al., 2020; Cuan-Baltazar et al., 2020; Nurbakova et al., 2020; Smith et al., 2020; Kouzy et al., 2020; Krause et al., 2020; Tasnim et al., 2020; Erku et al., 2020), but to the best of our knowledge, they assume that the initial message in a cascade is fake and do not study the mechanism of medical information distortion. We aim to answer two research questions: **RQ1**: *What are PF tweets on healthcare topics that generate information cascades?* **RQ2**: *How does a transformation of the initial tweet involve misinformation?*

Proceedings of the 3rd International Workshop on Rumours and Deception in Social Media (RDSM), pages 32–37
Barcelona, Spain (Online), December 12, 2020.

2 Materials and Methods for Detection of Topic Shift and Information Distortion

We collected 10M tweets in English about the controversy surrounding Covid-19 medical treatment published between 30/03/20-13/07/20 by querying Twitter API with the keywords, such as [*chloroquine, hydroxychloroquine, HCQ, Hydroxychloroquinum, azithromicyn, Raoult, remdesivir, tocilizumab*][1] (Noel et al., 2020). The data contained 141,866 original tweets, the rest are retweets[2]. As we focused on the analysis of information cascades ($2 * 10^4$), we only considered a subset of the dataset. First, we determined the *initial tweets of the cascades* among the union of 10^3 the most retweeted and 10^3 the most quoted tweets (1,356 unique tweet IDs). Then we added *cascades hops*, i.e. tweets with fields quoted_status.id or in_reply_to_status_id containing initial tweet IDs. The maximal cascade depth with the initial tweet in the resulting dataset is 10 (see examples in Fig.1). For further analysis, we considered the field text.

We analysed **topic shift** within information cascades by comparing (1) two neighbouring hops within a cascade $\Delta^{(i-1)}$ (Fig. 2(a)) and (2) each hop within a cascade with the initial tweet $\Delta^{(0)}$ (Fig. 2(b)). To analyse topic shift, we encode tweets with the state-of-the-art sentence embedding model USE (Universal Sentence Encoder) (Cer et al., 2018). Then, we computed the cosine similarity between USE embeddings (Singhal, 2001) and transformed it into distance by subtracting the obtained values from 1.

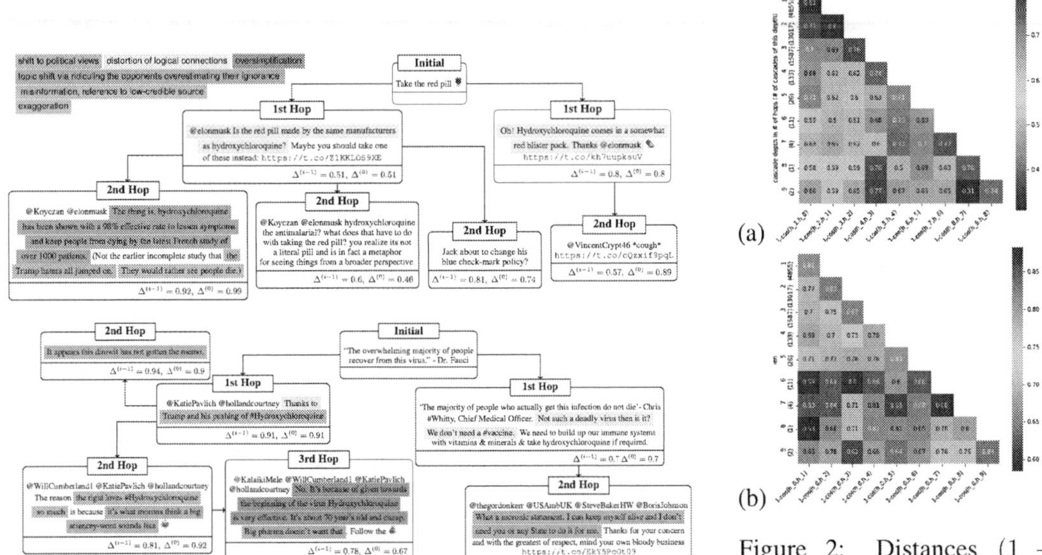

Figure 1: Examples of cascades and information distortion within them

Figure 2: Distances ($1 - cos$) between: (a) neighbouring hops, (b) hops and initial tweets

To identify information distortion types in cascades, we **manually performed semantic analysis** of tweet content. We examined key term distribution in cascades, explored their context in tweets and verified logical relations among medical terminology (see Table 1). The context analysis helped to recognise term substitutions and the substitution analysis to detect information distortion w.r.t. the initial tweet.

Our analysis also leans on **topic modelling**. We used Latent Dirichlet Allocation, LDA (Hoffman et al., 2010) from the `scikit-learn` tool. As tweets are short, we considered only the first topic. (Abd-Alrazaq et al., 2020) distinguish four main discussion themes on Twitter during the current pandemic: origin of the virus; its sources; its impact on people, countries, and the economy; and ways of mitigating the risk of infection. This set lacks medical disease description and ways to treat Covid-19 (symptoms, diagnosis, drugs, etc). Thus, we were also interested in **references to other disease related terms (DRT)** within cascades, as they can indicate distortion. To examine them, we extracted a list of hyponyms of the word *disease* from WordNet corpus accessed via `NLTK` library, to which we added terms like *plague,*

---

[1] The query was updated throughout the collection period based on new information about Covid-19 and its possible treatment.
[2] Some tweets attained 422K retweets, e.g. `https://bit.ly/32TPeSt` or `https://bit.ly/3gYDcfE`

Table 1: Disease related terms (DRTs) in cascades and their prevailing context

Visualisation of mentions of DRTs	Context and terms
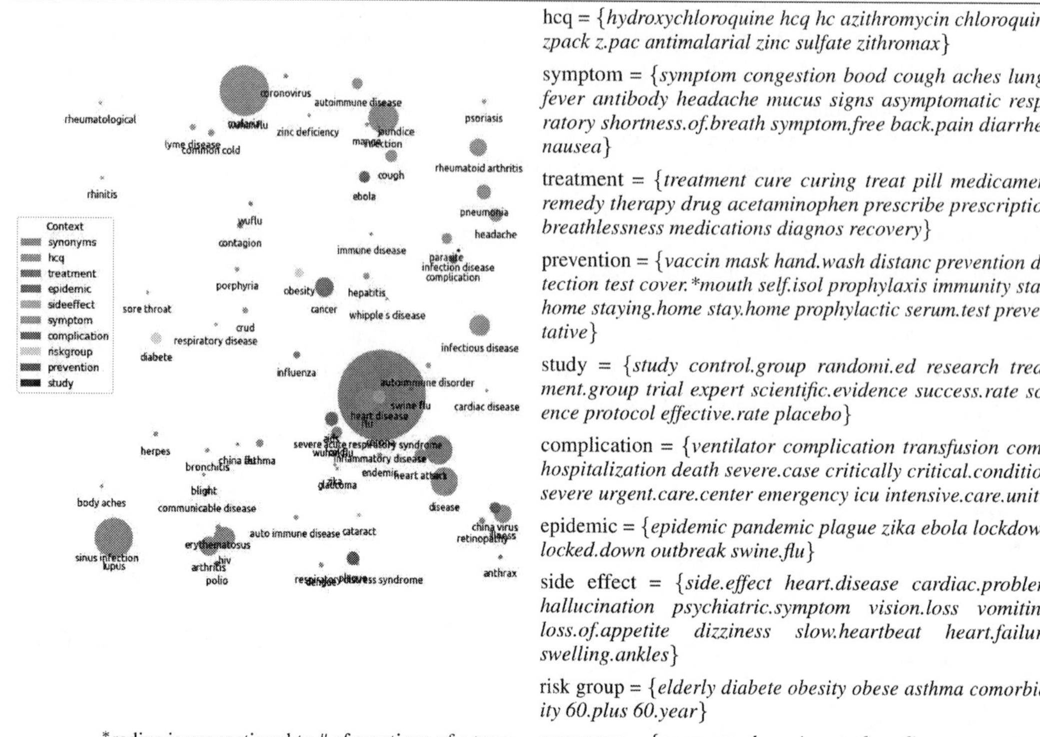 *radius is proportional to # of mentions of a term	hcq = {*hydroxychloroquine hcq hc azithromycin chloroquine zpack z.pac antimalarial zinc sulfate zithromax*} symptom = {*symptom congestion bood cough aches lungs fever antibody headache mucus signs asymptomatic respiratory shortness.of.breath symptom.free back.pain diarrhea nausea*} treatment = {*treatment cure curing treat pill medicament remedy therapy drug acetaminophen prescribe prescription breathlessness medications diagnos recovery*} prevention = {*vaccin mask hand.wash distanc prevention detection test cover.*mouth self.isol prophylaxis immunity stayhome staying.home stay.home prophylactic serum.test preventative*} study = {*study control.group randomi.ed research treatment.group trial expert scientific.evidence success.rate science protocol effective.rate placebo*} complication = {*ventilator complication transfusion coma hospitalization death severe.case critically critical.condition severe urgent.care.center emergency icu intensive.care.unit*} epidemic = {*epidemic pandemic plague zika ebola lockdown locked.down outbreak swine.flu*} side effect = {*side.effect heart.disease cardiac.problem hallucination psychiatric.symptom vision.loss vomiting loss.of.appetite dizziness slow.heartbeat heart.failure swelling.ankles*} risk group = {*elderly diabete obesity obese asthma comorbidity 60.plus 60.year*} synonyms = {*corona wuhan.virus wuhan.disease sars.cov.2 covid19 covid c19 coronovirus chinese.flu china.flu cv.19 sars.cov sars chinese.plague coronahoax wuhanflu*}

swine.flu, bird.flu, hiv, malaria, cough, wuhanflu, sars, cardiac.disease, china.flu, covid, coronovirus, cancer, obesity, diabete. We checked the appearance of these terms in the texts. In addition, we investigated the context in which these DRTs were mentioned such as: *hydroxychloroquine (HCQ), symptom, treatment, prevention, propagation, study, complication, epidemic, side effect, risk group, synonyms, plague reference, other issues*. Each context is defined by a set of terms (see Table 1). We then looked at the co-occurrence of DRTs and context terms in a tweet in order to predict the DRT context. This allowed us to gain a better understanding of topic shift related to references of other diseases.

3   Results

We identified the top-10 words characterising the first topic of each hop using LDA. We represented each hop as a binary vector built over the words of all hops. For visualisation, we applied Principle Component Analysis (PCA) (Tipping and Bishop, 1999) with two variables (see Fig. 3). Note that the first three hops are rather distant from the initial tweet, while the fourth hop is quite close to the initial tweet (its role is not clear yet). Based on our analysis, 3,939 out of 21,585 unique tweets of cascades contain DRTs. Table 1 summarises the frequency of the terms and the contexts in which they were primarily used[3]. HCQ is the most used context. It brings up DRTs such as *lupus, rheumatoid arthritis, malaria, heart attack, respiratory disease*, etc. Chloroquine often substituted its derivative HCQ, as in the E. Musk's cascade about the research of French microbiologist D. Raoult. The terms *corona* and *sars* are often used to refer to Covid-19. As for the *treatment* context, the most typical DRTs are *cancer, aids, influeza*. Note that a given term is often used in multiple contexts but here, we report the dominant one. Thus, HCQ was discussed in the cascades regardless of their initial tweet and the PF who initiated them.

---

[3]We intentionally excluded the term '*covid*' from the plot, as it is the main topic of the cascades (mentioned in 2,200 texts).

Information cascades are rarely evoked by *healthcare professionals* (HS) (identified by their profile information) since they are less active on Twitter than politicians. HS quotes are often misinterpreted and their research results are misrepresented. Though most of the cascades in which HS took part were initiated by *journalists*, HS tweets were able to **terminate an information cascade**[4] by providing relevant information and ending discussion. Cascade terminators often come from the professional community (Ziegelmeyer et al., 2010). In Fig. 2b, distances between initial tweets and the last hops of cascades show essential semantic differences revealing topic shifts. Hops in 'heads' of deep cascades are closer to their initial tweets than those in 'tails'. As distances between neighbouring hops show more similarity to each other than to their initial tweet (Fig. 2a),

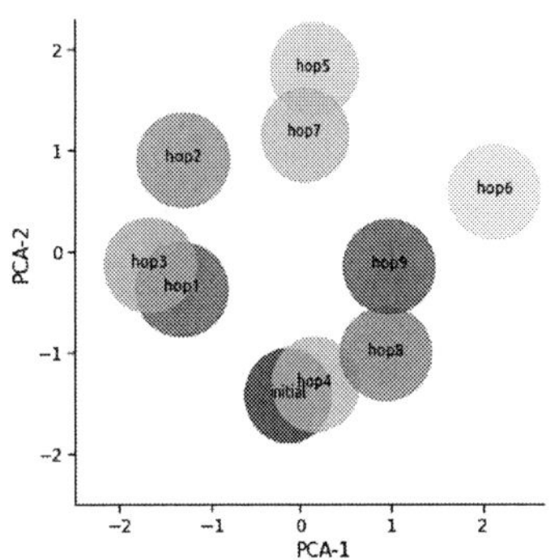

Figure 3: PCA on the first LDA topic

hops accumulate information mutability. The last hop is able to exhaust the cascade new topic.

Medical information is distorted via **erroneous logical conclusions** and mental operations of **oversimplification, overgeneralisation, exaggeration, substitution, omission of facts, insertion of erroneous conclusions, misuse of medical concepts** and **distortion of their connections**. The instances of the distortions occur in comments on the initial tweet and hops of cascades generated by the PF tweets. In fragments of cascades generated by comments on PF tweets in Fig. 1, distortion and **misinformation** appear due to **oversimplification** and **distortion of logical links**. **Omission of facts** is connected to **oversimplification**: HCQ efficacy in the Covid-19 treatment depends on patient anamnesis. **Misinformation** appears when a user did not provide a link to results of a French study he referred to while reacting on a red pill. The red pill meme reveals an unpleasant truth and is derived from a scene in the film *The Matrix*; an insertion of an erroneous conclusion occurred in comments where red pill is associated with HCQ. Ordinary users **exaggerate** consequences of government decisions. They **politicise** and **criminalise** these actions shifting the topic to political and business disputes (*RT @TribeforFreedom: Cuomo is dedicated to a vaccine (Bill Gates) So he does not allow the use of hydroxychloroquine*). **Overgeneralisation** often appears in references to a personal experience when a single fact was considered as a trend[5].

4   Conclusions

A topic shift is like the broken telephone effect (Boyd et al., 2010) when the message is altered during transmission, a typical cascade feature (Ribeiro et al., 2019). Thus, we showed that through this effect, PFs influence misinformation distribution on Twitter regardless of the quality of the information in their initial tweet. Users consider HSs as sources of 'raw information', which needs PF evaluation and approval. An interesting finding is that medical experts were able to stop the development of cascades by providing their factual and knowledgeable opinion. Intellectuals have the most influence on ordinary users' evaluation of the drugs efficacy research that is similar to the results of the cascades study in (Cha et al., 2010). We see the effect in the cascade evoked by comments on Musk's tweet. In the contexts of DRTs, we discovered the instances of medical information distortion. In contrast to previous works mainly focused on the initial spreading of fake news (Ahmed et al., 2020; Brennen et al., 2020), here we clarified the mechanism of the medical information distortion during the Covid-19 pandemic by analysing topic shifts within cascades. Usually, the medical topic is shifted to political and business disputes. We showed that cascade hops accumulate mutability of information. We found that after a noticeable topic shift occurring in the first 3 hops, there is a return to the original topic. Through context analysis, we improved the list of topics of (Abd-Alrazaq et al., 2020) adding those that are sensitive to medical information distortion. Our analysis provides valuable insights for the automatic detection and classification of medical information distortion.

---

[4]Example: *@eugenegu Hydroxychloroquine has known side effects including prolonging the heart QT interval (time between the Q wave and the T wave on an EKG), which is the time it takes for the ventricles to contract and relax. QT prolongation can cause Torsades de Pointes, a deadly heart rhythm*

[5]Example: *HYDROXYCHLOROQUINE cured my cousin and his wife, after 10 days of insurmountable suffering, in a matter of 24 hours...and he had existing heart issues. Did great*

References

A. Abd-Alrazaq, D. Alhuwail, M. Househ, M. Hamdi, and Z. Shah. 2020. Top Concerns of Tweeters During the COVID-19 Pandemic: Infoveillance Study. *J Med Internet Res*, 22(4):e19016, April.

W. Ahmed, J. Vidal-Alaball, J. Downing, and F.L. Seguí. 2020. COVID-19 and the 5G Conspiracy Theory: Social Network Analysis of Twitter Data. *Journal of Medical Internet Research*, 22(5):e19458.

D. Boyd, S. Golder, and G. Lotan. 2010. Tweet, Tweet, Retweet: Conversational Aspects of Retweeting on Twitter. In *2010 43rd Hawaii International Conference on System Sciences*, pages 1–10, January. ISSN: 1530-1605.

J.S. Brennen, F.M. Simon, P.N. Howard, and R.K. Nielsen. 2020. Types, Sources, and Claims of COVID-19 Misinformation. page 13.

Daniel Cer, Yinfei Yang, Sheng-yi Kong, Nan Hua, Nicole Limtiaco, Rhomni St. John, Noah Constant, Mario Guajardo-Cespedes, Steve Yuan, Chris Tar, Brian Strope, and Ray Kurzweil. 2018. Universal Sentence Encoder for English. In *Proceedings of the 2018 Conference on Empirical Methods in Natural Language Processing: System Demonstrations*, pages 169–174, Brussels, Belgium, November. Association for Computational Linguistics.

Meeyoung Cha, Hamed Haddadi, Fabrício Benevenuto, and P. Krishna Gummadi. 2010. Measuring user influence in twitter: The million follower fallacy.

Jose Yunam Cuan-Baltazar, Maria José Muñoz-Perez, Carolina Robledo-Vega, Maria Fernanda Pérez-Zepeda, and Elena Soto-Vega. 2020. Misinformation of COVID-19 on the Internet: Infodemiology Study. *JMIR Public Health and Surveillance*, 6(2):e18444. Company: JMIR Public Health and Surveillance Distributor: JMIR Public Health and Surveillance Institution: JMIR Public Health and Surveillance Label: JMIR Public Health and Surveillance Publisher: JMIR Publications Inc., Toronto, Canada.

Daniel A. Erku, Sewunet A. Belachew, Solomon Abrha, Mahipal Sinnollareddy, Jackson Thomas, Kathryn J. Steadman, and Wubshet H. Tesfaye. 2020. When fear and misinformation go viral: Pharmacists' role in deterring medication misinformation during the 'infodemic' surrounding COVID-19. *Research in Social and Administrative Pharmacy*.

Matthew D. Hoffman, David M. Blei, and Francis R. Bach. 2010. Online learning for latent dirichlet allocation. In John D. Lafferty, Christopher K. I. Williams, John Shawe-Taylor, Richard S. Zemel, and Aron Culotta, editors, *Advances in Neural Information Processing Systems 23: 24th Annual Conference on Neural Information Processing Systems 2010. Proceedings of a meeting held 6-9 December 2010, Vancouver, British Columbia, Canada*, pages 856–864. Curran Associates, Inc.

Ramez Kouzy, Joseph Abi Jaoude, Afif Kraitem, Molly B. El Alam, Basil Karam, Elio Adib, Jabra Zarka, Cindy Traboulsi, Elie W. Akl, and Khalil Baddour. 2020. Coronavirus Goes Viral: Quantifying the COVID-19 Misinformation Epidemic on Twitter. *Cureus*, 12(3), March. Publisher: Cureus Inc.

Nicole M. Krause, Isabelle Freiling, Becca Beets, and Dominique Brossard. 2020. Fact-checking as risk communication: the multi-layered risk of misinformation in times of COVID-19. *Journal of Risk Research*, April. Publisher: Routledge.

Marianne Noel, Liana Ermakova, Pedro Rammaciotti, Alexis Perrier, and Bilel Benbouzid. 2020. Controverse scientifique.

Diana Nurbakova, Liana Ermakova, and Irina Ovchinnikova. 2020. Understanding the Personality of Contributors to Information Cascades in Social Media in response to COVID-19 Pandemic. In *2020 International Conference on Data Mining Workshops, ICDM Workshops 2020, Sorrento, Italy, November 17-20, 2020*, page 8. IEEE.

B. Nye, J.J. Li, R. Patel, Y. Yang, I. Marshall, A. Nenkova, and B. Wallace. 2018. A Corpus with Multi-Level Annotations of Patients, Interventions and Outcomes to Support Language Processing for Medical Literature. In *Proceedings of the 56th Annual Meeting of the Association for Computational Linguistics (Volume 1: Long Papers)*, pages 197–207, Melbourne, Australia, July. Association for Computational Linguistics.

G. Pennycook and D.G. Rand. 2018. Lazy, Not Biased: Susceptibility to Partisan Fake News Is Better Explained by Lack of Reasoning Than by Motivated Reasoning.

G. Pennycook, J. McPhetres, Y. Zhang, J.G. Lu, and D.G. Rand. 2020. Fighting COVID-19 Misinformation on Social Media: Experimental Evidence for a Scalable Accuracy-Nudge Intervention. *Psychological Science*, 31(7):770–780, July. Publisher: SAGE Publications Inc.

Horta Manoel Ribeiro, Kristina Gligoric, and Robert West. 2019. Message distortion in information cascades. page 681–692.

D.M. Romero, W. Galuba, S. Asur, and B.A. Huberman. 2011. Influence and Passivity in Social Media. In Dimitrios Gunopulos, Thomas Hofmann, Donato Malerba, and Michalis Vazirgiannis, editors, *Machine Learning and Knowledge Discovery in Databases*, pages 18–33, Berlin, Heidelberg. Springer Berlin Heidelberg.

J. Shin, L. Jian, K. Driscoll, and F. Bar. 2018. The diffusion of misinformation on social media: Temporal pattern, message, and source. *Computers in Human Behavior*, 83:278–287, June.

Amit Singhal. 2001. Modern information retrieval: A brief overview. *IEEE Data Eng. Bull.*, 24:35–43.

Graeme D. Smith, Fowie Ng, and William Ho Cheung Li. 2020. COVID-19: Emerging compassion, courage and resilience in the face of misinformation and adversity. *Journal of Clinical Nursing*, 29(9-10):1425, May. Publisher: Wiley-Blackwell.

Samia Tasnim, Md Mahbub Hossain, and Hoimonty Mazumder. 2020. Impact of Rumors and Misinformation on COVID-19 in Social Media. *Journal of Preventive Medicine and Public Health*, 53(3):171–174. Publisher: The Korean Society for Preventive Medicine.

Michael E. Tipping and Christopher M. Bishop. 1999. Probabilistic principal component analysis. *Journal of the Royal Statistical Society: Series B (Statistical Methodology)*, 61(3):611–622.

A. Ziegelmeyer, F. Koessler, J. Bracht, and E. Winter. 2010. Fragility of information cascades: an experimental study using elicited beliefs. *Experimental Economics*, 13(2):121–145, June.

Revisiting Rumour Stance Classification: Dealing with Imbalanced Data

Yue Li and **Carolina Scarton**
Department of Computer Science, University of Sheffield, UK
riona.yueli@gmail.com, c.scarton@sheffield.ac.uk

Abstract

Correctly classifying stances of replies can be significantly helpful for the automatic detection and classification of online rumours. One major challenge is that there are considerably more non-relevant replies (*comments*) than informative ones (*supports* and *denies*), making the task highly imbalanced. In this paper we revisit the task of rumour stance classification, aiming to improve the performance over the informative minority classes. We experiment with traditional methods for imbalanced data treatment with feature- and BERT-based classifiers. Our models outperform all systems in RumourEval 2017 shared task and rank second in RumourEval 2019.

1 Introduction

A key step in the task of automatically analysing rumour veracity is to analyse the view of other users on a particular rumour (Procter et al., 2013), i.e. the stance of its replies. RumourEval 2017 and 2019 (Derczynski et al., 2017; Gorrell et al., 2019) are shared tasks that provide tree-structured conversation threads consisting of tweets directly or indirectly replying to a rumourous tweet and aim to label the stance of these replies towards the rumour (task A). Specifically, it is framed as a four-class classification problem: *support*, *deny*, *query*, and *comment* (SDQC). *Supports* and *denies* are arguably the most informative stances for rumour verification (Mendoza et al., 2010), while *comments* are considered the least useful. However, the data for this task is highly imbalanced with *supports* and *denies* corresponding, respectively, to 18% and 7% of instances in the RumourEval 2017, while *comments* are 66%.[1]

Systems submitted for RumourEval 2017 task A, evaluated in terms of accuracy, have a high performance for the majority class (*comments*), whilst the minority classes are under-performed. Mama Edha (García Lozano et al., 2017) adjusts the weights of the four labels, while only correctly classifying 1% *denies* and 37% *supports*. ECNU proposes a two-step classifier, however, only 1% of *denies* and 28% of *supports* are accurately predicted (Wang et al., 2017). IITP over-samples the underrepresented classes (Singh et al., 2017), although only 12% *denies* and 44% *supports* are recognised. The winner, Turing (Kochkina et al., 2017), is unable to identify any *denies* in the test data. In RumourEval 2019, with macro-$F1$ as evaluation metric, eventAI (Li et al., 2019) (third place) achieves 55% and 79% of correct *supports* and *denies*, respectively. Ranked first, BLCU NLP (Yang et al., 2019) increases the *supports* and *denies* with external similar datasets. However, the expanded data is still skewed towards *comments*, and 38% *supports* and 51% *denies* are accurately predicted. UPV (Ghanem et al., 2019) set different weights for each class, but 72% *supports* and 91% *denies* are mis-classified. Despite GWU (Hamidian and Diab, 2019) designing a rule-based model to help predict the instances of minority classes, their system does not correctly classify any *denies*. Other work (Zubiaga et al., 2018; Akhtar et al., 2018; Ma et al., 2018; Xuan and Xia, 2019) that also experiments with RumourEval or PHEME datasets does not consider imbalanced data approaches, and under-performs in *support* and *deny* classes.

In this paper, we experiment with well-known techniques for dealing with data imbalanced problems (e.g. SMOTE (Chawla et al., 2002)). We create feature-based models and a BERT-based model (Devlin

[1]RumourEval 2019 dataset has a similar distribution: *supports* $= 14\%$, *denies* $= 7\%$ and *comments* $= 72\%$.

Proceedings of the 3rd International Workshop on Rumours and Deception in Social Media (RDSM), pages 38–44
Barcelona, Spain (Online), December 12, 2020.

et al., 2019). Results show that our models not only have higher performance for the minority classes but also have higher overall performance (in terms of macro-$F1$ and geometric mean recall – GMR) than all systems submitted to RumourEval 2017 and all but one system submitted to RumourEval 2019.[2]

2 Resampling mechanisms and threshold-moving

Random under- or over-sampling (RUS or ROS)   RUS randomly discards samples in the majority class so that the class proportions can be balanced. Generally, it is computationally more efficient than over-sampling because it reduces the training data, although it may lead to under-fitting. ROS re-balances the class proportions through randomly replicating samples in the minority classes. However, these replications can increase the possibility of over-fitting (Prati et al., 2009).

Synthetic minority over-sampling technique (SMOTE)   is one of the most popular methods to over sample the minority class (Chawla et al., 2002). The mechanism is to artificially generate new samples based on k-nearest neighbours of each observation in the minority class. Although SMOTE also has other variants, such as Borderline-SMOTE (Han et al., 2005), we only experiment with the original SMOTE.

Adaptive synthetic (ADASYN) sampling approach   (He et al., 2008) is another over-sampling method similar to SMOTE, where the new synthetic samples are also interpolated based on each observation's k-nearest neighbours in the minority class. The main difference between SMOTE and ADASYN is the number of synthetic samples generated for each observation in the minority class. For SMOTE, it only depends on the required ratio of over-sampling, while in ADASYN, the number depends on the level of hardness of learning the data observation. ADASYN may focus too much on outliers, while SMOTE may associate outliers with inliers. Therefore, both of them could result in a sub-optimal decision.

SMOTE + Edited Nearest Neighbors (ENN) (SMOTEENN)   is a hybrid resampling method that combines over-sampling and under-sampling (Batista et al., 2004). Generally, it can achieve better performance than solely using SMOTE. In this method, firstly, the minority class is over-sampled by SMOTE. Then ENN will examine both majority and minority class and remove the data samples that are mis-classified by their three-nearest neighbours, which works as a data cleaning method.

Threshold-moving (TM)   (Maloof, 2003; Sheng and Ling, 2006) usually does not change the original class proportions. The classifier is trained with the imbalanced data, but the decision threshold that transforms the output probability into class label is changed. For example, we usually set 0.5 for a balanced binary classification. As there is no closed-form expression for a threshold that can maximise macro-$F1$ (Lipton et al., 2014), we set the threshold according to the class proportions, which has been proved to maximise macro accuracy based on two assumptions: (1) the class proportion of the test set is similar to that of the training set, and (2) the prior of a class is equivalent to its proportion in the training set (Collell et al., 2018). Therefore, our process for threshold moving is: (1) compute the output probability P_k for class k and (2) assign the class with highest P_k/a_k, $a_k = num_k/num_{total}$, in which num_k is the number of class k in the training set, and num_{total} is the total number of the training set.

3 Experiments and Results

3.1 Experimental setup

Techniques presented in Section 2 are explored in two types of classification models. We use implementations of resampling methods from the `imbalanced-learn` python toolkit (Lemaître et al., 2017).

Feature-based classifiers   Our feature-based approach is an adaptation of (Aker et al., 2017). We use Twitter-based features like number of re-tweets, presence of URLs and hashtags, number of followers for the user, among others. These features are then concatenated with a word vector representation for the tweets, using a pre-trained Twitter GloVe embedding model (Pennington et al., 2014). We train Random Forest (`RF`), Multi-Layer Perceptron (`MLP`) and Logistic Regression with stochastic gradient descent (`LR-SGD`) models using the `scikit-learn` python toolkit (Pedregosa et al., 2011).

---

[2]Our implementation is available at `https://github.com/YLi999/RumorStanceClassification`

	Average GMR and standard deviations						
	NT	**RUS**	**ROS**	**SMOTE**	**ADASYN**	**SMOTEEN**	**TM**
RF	0.000 ± 0.000	0.513 ± 0.025	0.453 ± 0.039	0.229 ± 0.242	0.000 ± 0.000	0.037 ± 0.079	0.457 ± 0.040
MLP	0.357 ± 0.139	0.541 ± 0.046	0.428 ± 0.154	0.442 ± 0.078	0.494 ± 0.057	0.477 ± 0.035	0.508 ± 0.036
LR-SGD	0.000 ± 0.000	0.519 ± 0.076	0.149 ± 0.195	0.234 ± 0.162	0.110 ± 0.178	0.230 ± 0.178	0.409 ± 0.067
BERT	0.482 ± 0.057	0.622 ± 0.027	0.442 ± 0.056	-	-	-	**0.626 ± 0.028**
	Average macro-F1 and standard deviations						
	NT	**RUS**	**ROS**	**SMOTE**	**ADASYN**	**SMOTEEN**	**TM**
RF	0.345 ± 0.012	0.509 ± 0.020	0.519 ± 0.011	0.529 ± 0.018	0.514 ± 0.002	0.333 ± 0.005	0.466 ± 0.014
MLP	0.466 ± 0.026	0.486 ± 0.038	0.507 ± 0.031	0.505 ± 0.044	0.531 ± 0.036	0.461 ± 0.024	0.502 ± 0.031
LR-SGD	0.402 ± 0.009	0.481 ± 0.065	0.384 ± 0.035	0.423 ± 0.024	0.418 ± 0.020	0.367 ± 0.020	0.534 ± 0.037
BERT	**0.584 ± 0.029**	0.502 ± 0.030	0.529 ± 0.025	-	-	-	0.540 ± 0.013

Table 1: GMR and macro-$F1$ on RumourEval 2017 development set. Best results overall are in bold. Best result for each approach are underlined.

BERT-based classifier (BERT)   We employ the pre-trained *BERT-base-uncased* model (Devlin et al., 2018) with 12 transformer layers, hidden unit size of 768, 12 attention heads, and 110M parameters. The inputs are the texts of a rumourous tweet and a reply tweet, and we fine tune for three epochs with a batch size of 16, using the `ktrain` (Maiya, 2020) toolkit. During training, we apply the 1 *cycle policy* (Smith, 2018), and search the optimal learning rate among $5e^{-5}$, $4e^{-5}$, $3e^{-5}$, and $2e^{-5}$. Since ROS, RUS and TM can be directly applied to raw text, we only apply BERT with these three methods.

Evaluation   For evaluation we use macro-$F1$ and GMR. Macro-$F1$ is the arithmetic mean between the $F1$-score $F_{1,c}$ of each class c: macro-$F1 = \frac{\sum_{i=1}^{C} F_{1,c}}{C}$ and is commonly applied in the evaluation of imbalanced binary classification. However, for multi-class problems, it is not robust to poor performance of the minority classes. GMR is denoted as $\sqrt[C]{\prod_{c=1}^{C} R_c}$, in which R_c is the recall of class c. False negatives may be more relevant than false positives in an imbalanced problem, therefore, it is important to assess models using recall-based metrics. Combining GMR with macro-$F1$ for evaluation can avoid choosing a model with high macro-$F1$ but actually with low recall for the minority classes.

3.2   Models assessment

For this experiment, we consider RumourEval 2017 data only.[3] We run each experiment 10 times to model variability and test on the development set. As baselines, we also train systems without any imbalanced data treatment (NT). Table 1 shows average and standard deviation of GMR and macro-$F1$. Although BERT in the NT case shows the highest macro-$F1 = 0.584$, BERT with TM (macro-$F1 = 0.540$) is still a better system, since it has the highest GMR $= 0.626$ and performs significantly better for *supports* and *denies*. When using feature-based classifiers, RUS leads to better models than the other approaches for both macro-$F1$ and GMR. The feature-based training data is high dimensional, which may harm the performance of resampling methods that are based on k-nearest neighbours. Systems with GMR $= 0$ are the worst case, since they could not correctly classify any *denies* in any of the 10 iterations (e.g. RF with ADASYN). Other systems, such as LR-SGD with SMOTE, fail to correctly predict any *denies* most of the time in 10 experiments, and consequently have high standard deviation of GMR (larger than 0.1). When using BERT, both RUS and TM result in a relatively good prediction on the minority classes, while TM perform better on the *comments* – the macro-$F1$ of BERT with TM is larger than that of BERT with RUS, although their GMRs are almost the same. TM works well with our neural models, BERT and MLP, which can provide good estimation of posterior probabilities. Finally, this analysis highlights the necessity of using both GMR and macro-$F1$ for evaluation. Some systems with high macro-$F1$ have low GMR, such as MLP with ADASYN (macro-$F1 = 0.531$, GMR $= 0.494$).

---

[3]Since RumourEval 2019 has Reddit data, it is not possible to use the same level of metadata available for tweets in this dataset, which justify our focus only on RumourEval 2017 data for model selection.

	GMR	macro-F1
BERT-TM(ensemble)	**0.635**	**0.536**
BERT-TM(single)	0.626	0.513
FBE-RUS	0.618	0.484
BERT-NT(single)	0.403	0.516
NileTMRG	0.363	0.452
ECNU	0.214	0.467
Turing	0.000	0.434

Table 2: Comparison with selected systems from RumourEval 2017.

	GMR	macro-F1
eventAI	**0.726**	0.578
BERT-TM(single)	0.618	0.561
BERT-TM(ensemble)	0.605	0.571
BLCU NLP	0.571	**0.619**
BUT-FIT	0.519	0.607

Table 3: Comparison with selected systems from RumourEval 2019.

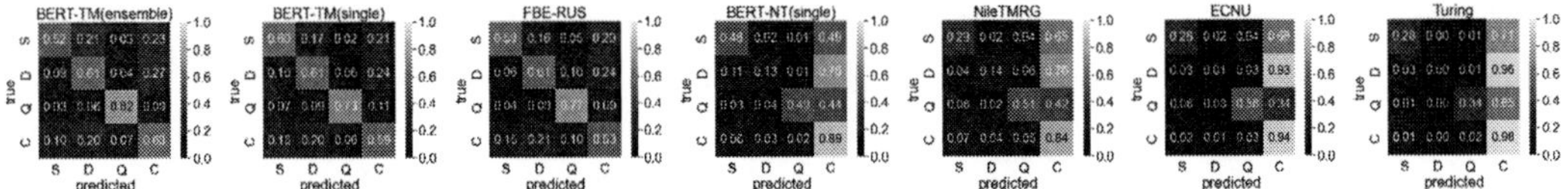

Figure 1: Confusion matrix for proposed models and selected systems for RumourEval 2017

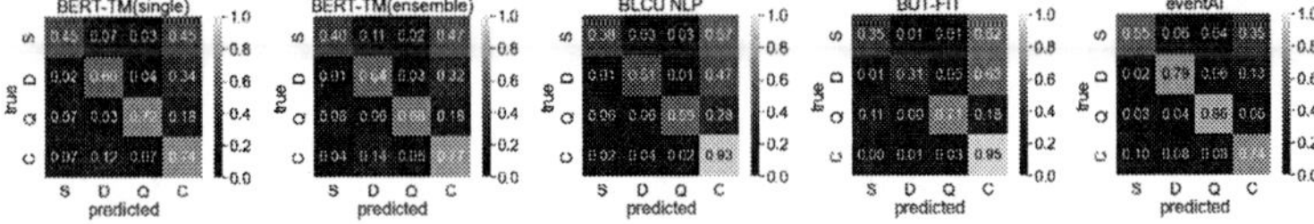

Figure 2: Confusion matrix for proposed models and selected systems for RumourEval 2019

3.3 Comparison with RumourEval submitted systems

As BERT with TM (BERT-TM(single)) is the best model on RumourEval 2017 development set, we further test it on both RumourEval 2017 and 2019 test sets, and compare it with the submitted systems. We also implement an ensemble of the three feature-based models (RF, MLP and LR-SGD) with RUS (FBE-RUS), and a bagging ensemble of BERT-TM(single) (Collell et al., 2018) – BERT-TM(ensemble). The process of training BERT-TM(ensemble) is: (1) generate n training sets with simple bootstrap sampling; (2) fine-tune n BERT base classifiers; (3) compute the average of n probabilistic predictions for each class; and, (4) perform TM. Similar to BERT-TM(single), the threshold is set according to the class proportion of training data. The optimal number of the base classifiers n is determined by the performance on development data ($n = 15$ in our case). We also present results for BERT without TM (BERT-NT(single)) on RumourEval 2017 test set.

For RumourEval 2017, we compare our models with Turing, ECNU, and NileTMRG (Enayet and El-Beltagy, 2017) (Table 2). BERT-TM(single), BERT-TM(ensemble), and FBE-RUS outperform other systems, showing similar performance for *supports* and *denies*. After applying TM on the output of BERT-NT (BERT-TM(single)), the performance on the minority classes is significantly enhanced (Figure 1). For RumourEval 2019, our models are compared with BLCU NLP, BUT-FIT (Fajcik et al., 2019), and eventAI (Table 3), outperforming BLCU NLP and BUT-FIT on *supports* and *denies* (Figure 2). Although eventAI performs better than our models, some details about its architecture are not provided in the paper and the code is not publicly available.

4 Conclusion and future work

We experiment with traditional imbalanced data techniques for the task of rumour stance classification and show that: (i) our models are capable of outperforming all systems in RumourEval 2017 and all but one system in RumourEval 2019 in terms of both macro-$F1$ and GMR scores, and (ii) a more in-depth evaluation is needed in order to correctly assess this task. Further improvements may be achieved by employing model-based imbalanced data techniques (e.g. by setting different weights for each class during training), which is left as future work.

Acknowledgments

This work was funded by the WeVerify project (EU H2020, grant agreement: 825297).

References

Ahmet Aker, Leon Derczynski, and Kalina Bontcheva. 2017. Simple open stance classification for rumour analysis. In *Proceedings of the International Conference Recent Advances in Natural Language Processing, RANLP 2017*, pages 31–39, Varna, Bulgaria, September. INCOMA Ltd.

Md Shad Akhtar, Asif Ekbal, Sunny Narayan, Vikram Singh, and Erik Cambria. 2018. No, that never happened!! investigating rumors on twitter. *IEEE Intelligent Systems*, 33(5):8–15.

Gustavo E. A. P. A. Batista, Ronaldo C. Prati, and Maria Carolina Monard. 2004. A study of the behavior of several methods for balancing machine learning training data. *ACM SIGKDD explorations newsletter*, 6(1):20–29.

Nitesh V. Chawla, Kevin W. Bowyer, Lawrence O. Hall, and W. Philip Kegelmeyer. 2002. SMOTE: synthetic minority over-sampling technique. *Journal of artificial intelligence research*, 16:321–357.

Guillem Collell, Drazen Prelec, and Kaustubh R. Patil. 2018. A simple plug-in bagging ensemble based on threshold-moving for classifying binary and multiclass imbalanced data. *Neurocomputing*, 275:330–340.

Leon Derczynski, Kalina Bontcheva, Maria Liakata, Rob Procter, Geraldine Wong Sak Hoi, and Arkaitz Zubiaga. 2017. SemEval-2017 task 8: RumourEval: Determining rumour veracity and support for rumours. In *Proceedings of the 11th International Workshop on Semantic Evaluation (SemEval-2017)*, pages 69–76, Vancouver, Canada, August. Association for Computational Linguistics.

Jacob Devlin, Ming-Wei Chang, Kenton Lee, and Kristina Toutanova. 2018. Bert: Pre-training of deep bidirectional transformers for language understanding. *arXiv preprint arXiv:1810.04805*.

Jacob Devlin, Ming-Wei Chang, Kenton Lee, and Kristina Toutanova. 2019. BERT: Pre-training of deep bidirectional transformers for language understanding. In *Proceedings of the 2019 Conference of the North American Chapter of the Association for Computational Linguistics: Human Language Technologies, Volume 1 (Long and Short Papers)*, pages 4171–4186, Minneapolis, Minnesota, June. Association for Computational Linguistics.

Omar Enayet and Samhaa R El-Beltagy. 2017. Niletmrg at semeval-2017 task 8: Determining rumour and veracity support for rumours on twitter. In *Proceedings of the 11th International Workshop on Semantic Evaluation (SemEval-2017)*, pages 470–474.

Martin Fajcik, Lukáš Burget, and Pavel Smrz. 2019. But-fit at semeval-2019 task 7: Determining the rumour stance with pre-trained deep bidirectional transformers. *arXiv preprint arXiv:1902.10126*.

Marianela García Lozano, Hanna Lilja, Edward Tjörnhammar, and Maja Karasalo. 2017. Mama edha at SemEval-2017 task 8: Stance classification with CNN and rules. In *Proceedings of the 11th International Workshop on Semantic Evaluation (SemEval-2017)*, pages 481–485, Vancouver, Canada, August. Association for Computational Linguistics.

Bilal Ghanem, Alessandra Teresa Cignarella, Cristina Bosco, Paolo Rosso, and Francisco Manuel Rangel Pardo. 2019. Upv-28-unito at semeval-2019 task 7: Exploiting post's nesting and syntax information for rumor stance classification. In *Proceedings of the 13th International Workshop on Semantic Evaluation*, pages 1125–1131.

Genevieve Gorrell, Elena Kochkina, Maria Liakata, Ahmet Aker, Arkaitz Zubiaga, Kalina Bontcheva, and Leon Derczynski. 2019. SemEval-2019 task 7: RumourEval, determining rumour veracity and support for rumours. In *Proceedings of the 13th International Workshop on Semantic Evaluation*, pages 845–854, Minneapolis, Minnesota, USA, June. Association for Computational Linguistics.

Sardar Hamidian and Mona Diab. 2019. Gwu nlp at semeval-2019 task 7: Hybrid pipeline for rumour veracity and stance classification on social media. In *Proceedings of the 13th International Workshop on Semantic Evaluation*, pages 1115–1119.

Hui Han, Wen-Yuan Wang, and Bing-Huan Mao. 2005. Borderline-SMOTE: a new over-sampling method in imbalanced data sets learning. In *Proceedings of the International Conference on Intelligent Computing (ICIC 2005)*, pages 878–887, Hefei, China. Lecture Notes in Computer Science.

Haibo He, Yang Bai, Edwardo A Garcia, and Shutao Li. 2008. ADASYN: Adaptive synthetic sampling approach for imbalanced learning. In *Proceedings of the 2008 IEEE international joint conference on neural networks (IEEE world congress on computational intelligence)*, pages 1322–1328, Hong Kong, China. IEEE.

Elena Kochkina, Maria Liakata, and Isabelle Augenstein. 2017. Turing at SemEval-2017 task 8: Sequential approach to rumour stance classification with branch-LSTM. In *Proceedings of the 11th International Workshop on Semantic Evaluation (SemEval-2017)*, pages 475–480, Vancouver, Canada, August. Association for Computational Linguistics.

Guillaume Lemaître, Fernando Nogueira, and Christos K. Aridas. 2017. Imbalanced-learn: A python toolbox to tackle the curse of imbalanced datasets in machine learning. *Journal of Machine Learning Research*, 18(17):1–5.

Quanzhi Li, Qiong Zhang, and Luo Si. 2019. eventAI at SemEval-2019 task 7: Rumor detection on social media by exploiting content, user credibility and propagation information. In *Proceedings of the 13th International Workshop on Semantic Evaluation*, pages 855–859, Minneapolis, Minnesota, USA, June. Association for Computational Linguistics.

Zachary C. Lipton, Charles Elkan, and Balakrishnan Naryanaswamy. 2014. Optimal thresholding of classifiers to maximize F1 measure. *Mach Learn Knowl Discov Databases*, 8725:225–239.

Jing Ma, Wei Gao, and Kam-Fai Wong. 2018. Detect rumor and stance jointly by neural multi-task learning. In *Companion Proceedings of the The Web Conference 2018*, pages 585–593.

Arun S. Maiya. 2020. ktrain: A Low-Code Library for Augmented Machine Learning. *arXiv preprint arXiv:2004.10703*.

Marcus A Maloof. 2003. Learning when data sets are imbalanced and when costs are unequal and unknown. In *Proceedings of the ICML-2003 Workshop on Learning from Imbalanced Data Sets II*, Washington, DC, USA.

Marcelo Mendoza, Barbara Poblete, and Carlos Castillo. 2010. Twitter under crisis: can we trust what we RT? In *Proceedings of the First Workshop on Social Media Analytics*, pages 71–79, Washington, DC, USA. Association for Computing Machinery.

Fabian Pedregosa, Gaël Varoquaux, Alexandre Gramfort, Vincent Michel, Bertrand Thirion, Olivier Grisel, Mathieu Blondel, Peter Prettenhofer, Ron Weiss, Vincent Dubourg, Jake Vanderplas, Alexandre Passos, David Cournapeau, Matthieu Brucher, Matthieu Perrot, and Édouard Duchesnay. 2011. Scikit-learn: Machine Learning in Python. *Journal of Machine Learning Research*, 12:2825–2830.

Jeffrey Pennington, Richard Socher, and Christopher Manning. 2014. Glove: Global vectors for word representation. In *Proceedings of the 2014 Conference on Empirical Methods in Natural Language Processing (EMNLP)*, pages 1532–1543, Doha, Qatar, October. Association for Computational Linguistics.

Ronaldo C. Prati, Gustavo E. A. P. A. Batista, and Maria Carolina Monard. 2009. Data mining with imbalanced class distributions: concepts and methods. In *Proceedings of the 4th Indian International Conference on Artificial Intelligence (IICAI 2009)*, pages 359–376, Tumkur, Karnataka, India.

Rob Procter, Farida Vis, and Alex Voss. 2013. Reading the riots on twitter: methodological innovation for the analysis of big data. *International journal of social research methodology*, 16(3):197–214.

Victor S. Sheng and Charles X. Ling. 2006. Thresholding for making classifiers cost-sensitive. In *Proceedings of the Twenty-first National Conference on Artificial Intelligence (AAAI-06)*, pages 476–481, Boston, Massachusetts. American Association for Artificial Intelligence.

Vikram Singh, Sunny Narayan, Md Shad Akhtar, Asif Ekbal, and Pushpak Bhattacharyya. 2017. IITP at SemEval-2017 task 8 : A supervised approach for rumour evaluation. In *Proceedings of the 11th International Workshop on Semantic Evaluation (SemEval-2017)*, pages 497–501, Vancouver, Canada, August. Association for Computational Linguistics.

Leslie N. Smith. 2018. *A disciplined approach to neural network hyper-parameters: Part 1–learning rate, batch size, momentum, and weight decay*. US Naval Research Laboratory Technical Report 5510-026.

Feixiang Wang, Man Lan, and Yuanbin Wu. 2017. ECNU at SemEval-2017 task 8: Rumour evaluation using effective features and supervised ensemble models. In *Proceedings of the 11th International Workshop on Semantic Evaluation (SemEval-2017)*, pages 491–496, Vancouver, Canada, August. Association for Computational Linguistics.

Kaizhou Xuan and Rui Xia. 2019. Rumor stance classification via machine learning with text, user and propagation features. In *2019 International Conference on Data Mining Workshops (ICDMW)*, pages 560–566. IEEE.

Ruoyao Yang, Wanying Xie, Chunhua Liu, and Dong Yu. 2019. Blcu_nlp at semeval-2019 task 7: An inference chain-based gpt model for rumour evaluation. In *Proceedings of the 13th International Workshop on Semantic Evaluation*, pages 1090–1096.

Arkaitz Zubiaga, Elena Kochkina, Maria Liakata, Rob Procter, Michal Lukasik, Kalina Bontcheva, Trevor Cohn, and Isabelle Augenstein. 2018. Discourse-aware rumour stance classification in social media using sequential classifiers. *Information Processing & Management*, 54(2):273–290.

Fake news detection for the Russian language

Gleb Kuzmin[*]
Moscow Institute of Physics and Technology
/ Dolgoprudny, Russia
kuzmin.gyu@phystech.edu

Daniil Larionov[*]
Federal Research Center
"Computer Science and Control"
/ Moscow, Russia
dslarionov@isa.ru

Dina Pisarevskaya[*]
Federal Research Center
"Computer Science and Control"
/ Moscow, Russia
dinabpr@gmail.com

Ivan Smirnov
Federal Research Center
"Computer Science and Control"
/ Moscow, Russia
ivs@isa.ru

Abstract

In this paper, we trained and compared different models for fake news detection in Russian. For this task, we used such language features as bag-of-n-grams and bag of Rhetorical Structure Theory features, and BERT embeddings. We also compared the score of our models with the human score on this task and showed that our models deal with fake news detection better. We investigated the nature of fake news by dividing it into two non-overlapping classes: satire and fake news. As a result, we obtained the set of models for fake news detection; the best of these models achieved 0.889 F1-score on the test set for 2 classes and 0.9076 F1-score on 3 classes task.

1 Introduction

Fake news detection becomes a more significant task. It is connected with an increasing number of news in media and social media. To prevent rumors and misinformation from spreading in this news, we need to have a system able to detect fake news. It also could be useful because it's hard for people to recognize fake news. Approaches to this task are being developed for English. However, fake news texts can be written originally in different 'source' languages, i.e. in Russian. To tackle such content in social media, multilingual systems might be used. For Russian, only preliminary research for automated fake news detection was undertaken before. Moreover, fake news detection in Russian becomes more actual due to the appearance of new laws in the Russian legal system. For example, since 2020 the Russian legal system has special criminal law[1] for spreading fake news about emergencies. Furthermore, due to an increasing number of fake news, connected with Russia, it will be useful to have the fake news detection module for the Russian language to check original news in Russian.

In our paper, we trained and compared different models for fake news detection in Russian. We checked if various language features alone can be helpful for this task. As baseline models, we used Support Vector Machines (SVM) and Logistic Regression over a bag-of-n-grams. Also, we trained similar models, but over features obtained from the discourse parsing of news (Rhetorical Structure Theory (RST) discourse features, as in (Mann and Thompson, 1988)). For each news text, these features were constructed from its hierarchical discourse tree representation. It contains discourse (rhetorical) relations between text segments – discourse units, starting with the smallest 'leaf' segments - elementary discourse units. As the third model, we fine-tuned BERT (for the Russian language) for the fake news detection task.

---

[*] - These authors contributed equally.

[1] http://duma.gov.ru/news/29982/ (in Russian)

Proceedings of the 3rd International Workshop on Rumours and Deception in Social Media (RDSM), pages 45–57
Barcelona, Spain (Online), December 12, 2020.

We investigated all existing datasets for automated fake news detection for Russian. During our study, we also examined some hypotheses about different types of fake news. We divided our data into three parts - non-satirical fake news (that are equal to simple fake news), satirical fake news, and real news. The satirical fake news is the deceptive news, that was written deliberately with humorous purpose or at least without disinformation purpose. It is presented in the format and style of legitimate news articles (Rubin et al., 2016; De Sarkar et al., 2018). A more detailed description of the data splitting can be found in section 3. We established that, for the classification task, satirical news should be singled out as a separate class, among fake news and real news. Finally, we annotated the part of the dataset manually and compared the results of our models with human performance, setting goals for future research steps.

2   Related Work

1. Linguistic features. For English, linguistic features for fake news detection were studied during recent years. Content features are used in (Rashkin et al., 2017; Volkova et al., 2017; Ruchansky et al., 2017; Ma et al., 2018; Kochkina et al., 2018; Khattar et al., 2019). Text-based only approach is also used in (Ajao et al., 2019) (sentiment features), (Dungs et al., 2018; Wu et al., 2019) (stance detection). (Baly et al., 2018) suggest linguistic features (POS tags, sentiment scores, readability and subjectivity features, number of cognitive process words etc.) as well as source credibility features. (Karadzhov et al., 2017; Pérez-Rosas et al., 2018; Potthast et al., 2018) explore linguistic features, including ngrams, POS tags, readability and complexity features; psycholinguistic features from LIWC (Rashkin et al., 2017; Pérez-Rosas et al., 2018) and other sources (Rashkin et al., 2017), syntax features (Pérez-Rosas et al., 2018). There are several datasets, i.e. the dataset proposed in (Rashkin et al., 2017).

As for discourse features, (Karimi and Tang, 2019) incorporate hierarchical discourse-level structures for fake news detection. Structure-related properties (number of leaf nodes, preorder difference, parent-child distance) identify structural differences between fake and real news texts. The approach, based on discourse dependency trees, yields better results (82.19%) than approaches based on n-grams, on LIWC features, on RST discourse features taking the hierarchical structure of document into account (as in (Rubin and Lukoianova, 2015)), or BiGRNN-CNN and LSTM approaches using word embeddings in sentences. (Atanasova et al., 2019) propose a set of various features for context and discourse modeling. There are also discourse features among them, based on automated discourse parsing according to RST. They are focused on the direct relationship between a target sentence and other sentences in a segment and on the internal structure of a target sentence (number of nuclei and satellites). Such rhetorical relation types as Background, Enablement, Elaboration, Attribution are associated with factually-true examples.

Linguistic features for fake news detection have limitations and can be used in addition to automated fact-checking. I.e., Shuster (Schuster et al., 2020) study linguistic (stylistic) features for machine-generated misinformation detection and conclude that they are limited in detecting if texts generated by language models are fake, as such texts hide stylistic differences between falsified and truthful content.

Automated satire and biased text detection are closely connected tasks. For satire detection, absurdity feature (unexpected introduction of new named entities within the final sentence) (Rubin et al., 2016), POS features (Rubin et al., 2016; Yang et al., 2017; De Sarkar et al., 2018), psycholinguistic, readability and structural text features (Yang et al., 2017), sentiment scores and named entity features (De Sarkar et al., 2018) can be used. Satire can be distinguished from fake news using semantic representation with the BERT language model and with linguistic features based on textual coherence metrics that also include basic language features, such as readability features, sentence length, number of words (Levi et al., 2019). Language features, without checking external sources of information, are also mainly used for biased language and propaganda detection (Potthast et al., 2018; Da San Martino et al., 2019)

2. Claims verification. In automated fact-checking for English, (Thorne et al., 2018; Nie et al., 2019; Augenstein et al., 2019; Zhong et al., 2020; Portelli et al., 2020; Kochkina and Liakata, 2020) consider evidence detection and claims verification. Several recent studies are focused on evidence detection for explainable claim verification, based on semantic entailments for claims (Hanselowski et al., 2018; Ma et al., 2019), incorporating semantic similarity between comments and claims (Wu et al., 2020). BERT language model (Devlin et al., 2019) can be used for checking if a comment is factual (Stammbach et al., 2019), for retrieving evidence sentences and verifying the truthfulness of claims against the retrieved evidence sentences (Soleimani et al., 2020); there are also experiments on using it solely, without any external knowledge or explicit retrieval components (Lee et al., 2020). There are several datasets, i.e. PHEME (Zubiaga et al., 2016), LIAR (Wang, 2017), RumourEval (Derczynski et al., 2017), FEVER (Thorne et al., 2018).

For Russian, few initial research studies on fake news detection have been conducted, each one based on a single dataset. Basic lexical, syntactic, and discourse parameters were examined in (Pisarevskaya, 2017). The impact of named entities, verbs, and numbers was investigated in (Zaynutdinova et al., 2019). There are only a few datasets for fake news language and no existing datasets for claims verification. To build a misinformation detection system for social media posts in Russian, firstly we study language features for fake news detection. As discourse features were already regarded for fake news detection for Russian, we check their impact in our study too.

3   Data

3.1   Data Collection

To the best of our knowledge, there are three available datasets for fake news detection in Russian.

- 174 texts from (Pisarevskaya, 2017), with an equal number of fake and truthful texts, parsed in 2015-2017 from Russian news sources. The dataset is available upon request.

- texts from (Zaynutdinova et al., 2019). Total 8867 texts, with 1366 fakes and 7501 real ones. The dataset is also available upon request.

- Fake news dataset from the satire and fake news website `https://panorama.pub/`. This dataset is a part of the Taiga corpus for Russian, it is freely available at `https://tatianashavrina.github.io/taiga_site/downloads`. We have taken 1803 satirical texts.

We incorporate them for our research and base it on them. In our dataset, we used only one source of satirical news because we found only one reliable source of satirical fake news. It could induce some bias during model training. On the other hand, news from this source are written by different authors on various topics, so we suppose that one source could be enough.

3.2   Data Description

We created 5 smaller datasets from the described data and used each of them for model training. They are structured as follows:

1. train and test parts - non-satirical fake news and real news (Fakes & Fakes) (9041 samples, test size is 20 % of the dataset);

2. train and test parts - satirical fake news and real news (Satira Fakes & Satira Fakes) (10136 samples, test size is 20 % of the dataset with fixed seed);

3. train part - satirical fake news and real news, test part - non-satirical fake news and real news (Satira Fakes & Fakes) (9476 samples, fixed test size with 174 samples);

4. train part - satirical and non-satirical fake news and real news, test part - non-satirical fake news and real news (Fakes + Satira Fakes & Fakes) (11676 samples, fixed test size with 174 samples);

5. train and test part - satirical and non-satirical fake news and real news, 3 class classification (Fakes + Satira Fakes & Fakes + Satira Fakes) (11676 samples, test size is 20 % of the dataset with fixed seed).

In datasets 1-4 we were solving a binary classification problem, therefore, satirical and non-satirical fake news was treated as one class. In the 5th dataset, we considered the multiclass classification, therefore satirical and non-satirical fakes were treated as two different classes. We specially created 5 datasets to check some hypotheses on fake news structure. First of all, we would like to test if satirical fake news is different from non-satirical fake news. To test this hypothesis, we used datasets 2-5, while the first dataset contains only non-satirical fake news and real news and serves as a reference score for comparison.

4 Experiments

4.1 Baseline

We chose a basic method of a bag-of-n-grams, with TF-IDF preprocessing, for the baseline models. Pre-processing consists of removing control characters, removing http-like links, and optional lemmatization. Also, we chose to select a subset of the most informative features before training the model. This was done by computing ANOVA F-value for each feature and selecting **k** highest scored features, where **k** is a hyperparameter.

Within this framework, we trained a classification model, based on Support Vector Machines with RBF kernel. Also, for a 3-class classification task (Fakes + Satira Fakes & Fakes + Satira Fakes) we trained a Logistic Regression based model for better model interpretability. We employed Bayesian hyperparameter optimization (Snoek et al., 2012) with Hyperband (Li et al., 2016) early termination algorithm, in order to estimate optimal hyperparameter values (Table 8).

4.2 RST Features

For more advanced models, we employed features obtained from the RST parsing of our texts. We used the automated discourse parser for Russian firstly proposed in (Shelmanov et al., 2019). The first approach is a so-called "bag-of-rst" features. For each text, we have taken all the RST relations for all discourse units in the texts and encoded them into a one-hot vector. Such vectors were concatenated with feature vectors from the baseline model and used with an SVM-based classifier (Logistic Regression-based for the 3-class case). We employed the aforementioned hyperparameter optimization algorithm for this pipeline as well.

The second approach, aimed to better consider the hierarchical structure, was based on averaging embeddings of different discourse units. Firstly, we grouped nodes of the discourse tree by their relation. Then texts from both leaves of each node were concatenated and passed through the BERT-based sentence embedding model from (Kuratov and Arkhipov, 2019). Resulting vectors were averaged for each type of RST relation (see Table 1). We used the concatenation of averaged vectors together with the SVM based classifier.

Relation		
attribution	background	cause-effect
concession	condition	contrast
elaboration	interpretation-evaluation	joint
preparation	purpose	same-unit
solutionhood		

Table 1: List of RST relations extracted by (Shelmanov et al., 2019)

However, the model from the second approach was not able to learn to predict anything from this set of features. It looks like the model always chooses to predict a single class i.e. works as a constant predictor. Thus, we chose to focus on bag-of-rst and BERT based models.

4.3 Feature Importance

For baseline and "bag-of-rst" models we extracted feature importance using Shapley Additive explanations method from (Ribeiro et al., 2016). We can see on charts (Figure 1) that the most important feature

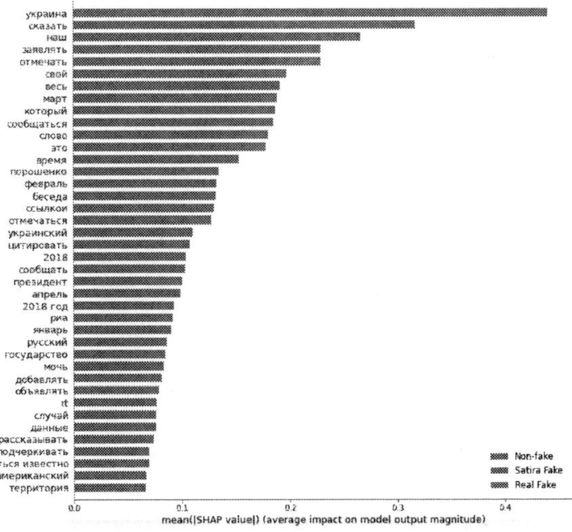

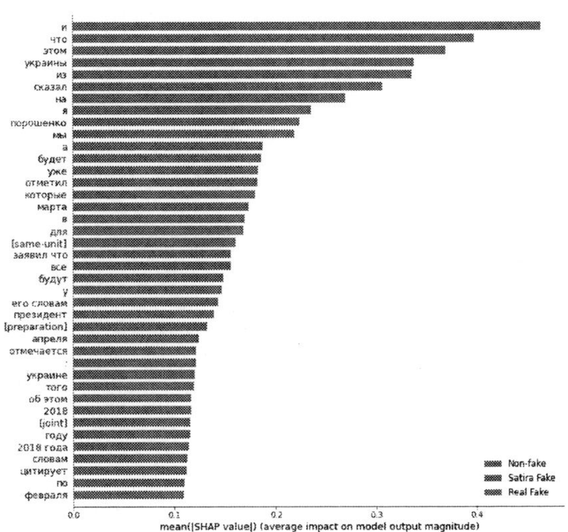

Figure 1: The most important features for the Logistic Regression-based baseline model and the "bag-of-rst" model

is the word "Ukraine". This is because the Fakes part of our dataset is hugely based on Ukraine-related texts about the Russia-Ukraine conflict. Thus, this is a rich area for fakes generation by both sides of the conflict. Also, another Ukraine-related feature in the list is "Poroshenko", the surname of the former Ukrainian president Petro Poroshenko. Another conclusion that can be drawn from these features is that RST relations are among the most important features for the "bag-of-rst" classifier. And these relations impacts all classes, the classifier does not overfit on them but uses these features together with bag-of-n-grams features. Speaking on RST relations, we see several examples: the presence of "same-unit" or "preparation" relations almost always moves a model prediction towards "satire" class, while "joint" never does so. "Same-unit" is less informative as an utility relation (Shelmanov et al., 2019).

4.4 Ensembles

To further improve the quality of our models, we decided to the evaluate ensembles of best models. Firstly, we chose to implement an ensemble of baseline and 'bag-of-rst' models. The weight coefficients were additionally optimized for maximum performance. However, the ensemble only slightly outperforms the 'bag-of-rst' model by 0.001-0.002 of F1-score. At the same time, the ensemble of BERT models shows F1-score 0.02 less than just the best BERT model.

4.5 Fine-tuning BERT

In the last few years, BERT-based models showed state-of-the-art results in various NLP tasks. Particularly, these models also showed good results in sequence classification tasks. As one of the models, we used pre-trained RuBERT[2] from DeepPavlov (Burtsev et al., 2018) with Hugging Face (Wolf et al., 2019). In the process of fine-tuning, we trained only the last fully-connected layer with weighted cross-entropy as the loss function. That was done due to the unbalanced class distribution in our data. For tuning, we also used Adam optimizer with the linear scheduler (decays the learning rate of optimizer by γ each γ-steps). As we used BERT for tuning, all our news texts were truncated at a size of 512 tokens. Nevertheless, we found that 512 tokens are enough in the case of our dataset because only 1% of news has a length of more than 512 tokens. More detailed model parameters are described in the results section alongside its performance.

5 Results

5.1 BERT

For the model trained on datasets 1-4 we used the following model parameters: batch size - 8, epochs - 20, lr - $9.2 \cdot 10^{-5}$, max_tokens - 512, γ - 0.357, γ-steps - 9. For 3 classes classification on dataset 5 we used other parameters: batch size - 8, epochs - 20, lr - $9.98 \cdot 10^{-5}$, max_tokens - 512, γ - 0.436, γ-steps - 8 (Table 2).

Dataset	F1-score, train/test	accuracy, train/test	roc-auc, train/test
1	0.765/0.778	0.883/0.890	0.745/0.752
2	0.881/0.887	0.906/0.909	0.891/0.895
3	0.446/0.333	0.806/0.500	0.500/0.500
4	0.715/0.546	0.738/0.546	0.718/0.546
5	0.741/0.748	0.823/0.822	0.913/0.909

Table 2: RuBERT fine-tuning results

5.2 Baseline and RST Features Results

After stages of bayesian optimization, we found the final sets of hyperparameters (Table 9).

We achieved decent results both on binary classification datasets (1-4) and 3-class cases (5). As we can see in Table 3, RST features do not improve the performance of bag-of-n-grams models. However, as we see in Figure 1, the RST-based model has RST features in the top-20 of the most important features, thus such a model learns differently and uses the discourse structure for text scoring.

It is worth mentioning that we discovered strong negative correlation between the f1-score of the model and the minimal n-gram size, used for tokenization. That's why we always have an n-gram range of (1, ..). Also, all the top features on all the models are unigrams (see Figure 1). That's why we state that fake texts may be characterized by a very specific vocabulary and discourse structure, used by their authors.

---

[2]`https://huggingface.co/DeepPavlov/rubert-base-cased-sentence`

Dataset	SVM-baseline	SVM "bag-of-rst"	LogReg-baseline	LogReg "bag-of-rst"
1	0.8800	0.8796	0.8875	0.8829
2	0.9576	0.9509	0.9513	0.9562
3	0.5950	0.5886	0.5919	0.5944
4	0.5600	0.5671	0.5576	0.5743
5	0.9084	0.8901	0.9076	0.9042

Table 3: F1-scores for baseline and RST feature-based models results for binary tasks and 3-class case

5.3 Difference between Satirical Fakes and Fakes

Our initial hypothesis was that satire is similar to fake news, thus it can be used to improve classification performance. Indeed, satire texts are much easier to obtain, because they are typically hosted on well-known satire-news media. And people are equivalently likely to misinterpret satire texts as truthful ones. However, our experimental evidence shows that satire significantly differs from real fakes. According to Table 2 and Table 3, the performance on the the dataset 4, where satire and real fakes are mixed, is worse than on the dataset 3, where the model is trained on satire texts and tested against real fakes. Thus, we decided to separate satire texts, non-fake texts, and Fakes into 3 different classes (dataset 5).

5.4 Comparison with Human Performance

In this section, we compared the performance of our models with the human score. Although the datasets were already annotated before publishing them, for this task we manually labeled, in addition, about 500 random texts from the test part of our dataset for 3 class classification, to investigate the results in detail. Texts were labeled by 3 annotators, the guidelines were short and not explicit: we aimed to check how human annotators, not experts, define if news texts are fake, satirical, or real. It is worth noting that we already had a ground truth annotation for our datasets. So this additional manual annotation was only used for comparison of our models with the human score on the part of the test set, and for checking the cases, where the models gave wrong predictions, more thoroughly.

Then we used this manually labeled data as an additional (separate) test set for our models. The results are shown in Table 4.

Model	F1-score	accuracy
1 annotator	0.564	0.731
2 annotator	0.516	0.705
3 annotator	0.806	0.881
RuBERT	0.740	0.815
Best model (SVM)	0.908	0.941

Table 4: Comparison of our models with human performance

6 Error Analysis and Discussion

As an automated fake news detection system could be used as a preliminary filter before human evaluation, it is important to reduce the number of false positives for fakes and satire. For the selected BERT model, the false positives rate is only 0.05 for fakes and 0.06 for satire on the test set (0.05 and 0.05 on its human-annotated part). For the n-grams model, the false positives rate is 0.01 for fakes and 0.02 for satire on the test set (0.02 and 0.03 on its human-annotated part).

We performed an error analysis on the annotated part of the test set, to compare the model's results with human assessments. Metrics on this part slightly differ from the metrics on the whole test set (Table 5). Human annotators were also unsure of labeling satirical texts that were mispredicted as fake news. Real texts, that were mispredicted by the model as fake news, were mostly labeled correctly by the annotators but contained loaded language, emotional lexicon. 70% of them were about the politics of Ukraine (see

Class	Precision (whole set)	Precision (annotated)	Recall (whole set)	Recall (annotated)
Real news	0.867	0.849	0.895	0.908
Fake news	0.635	0.638	0.544	0.507
Satirical news	0.774	0.806	0.768	0.755

Table 5: BERT metrics on the test set

Section 4.3), yielding that dataset topics should have been more diverse. Real texts mispredicted as satire were about unusual events and required some additional knowledge, all of them were predicted correctly by the annotators with 100% inter-annotator consistency to satirical texts and, to a greater degree, fake texts that were mispredicted as real news, in some cases, it is hard to understand why they are not real without real-world knowledge, and annotators provide different labels. In these cases, people also can be mistaken.

We also looked at human labels and n-grams model results on the annotated part of the test set (Table 6). In cases where the n-grams model mispredicted fake news as satirical news and satirical news as fake news, human annotators also did not reach agreement: the texts were about politics and required the knowledge of facts. For all fake texts that were labeled as real by the model, the majority vote of annotators would also provide the 'fake' tag.

Class	Precision	Recall
Real news	0.932	0.956
Fake news	0.867	0.712
Satirical news	0.896	0.936

Table 6: N-grams model metrics on the annotated test set part

We investigated 'gold' labels and labels created by annotators and figured out possible issues of concern.

Only one annotator provided tags that were close to the 'gold' labels (f1 score 0.806). While checking the inter-annotator agreement between 3 annotators (about 500 texts), we found out that the annotators reached a substantial agreement only in distinguishing real news from fake and satirical news (Table 7).

According to the majority vote, it is more simple to detect satire. 71% satirical texts were annotated correctly, in comparison with 25% fake texts. Fake texts are mixed up with real news: in 2% cases, fake texts were labeled as satire by one single annotator, in other cases, they were labeled as fake or real ones. It also yields the subjectivity of a manual approach to fake news detection.

Agreement	Fleiss' kappa
3 classes: satirical, fake, real news	0.485
2 classes: 1) fake and real news 2) satirical news	0.553
2 classes: 1) real news 2) fake and satirical news	0.629

Table 7: Inter-annotator agreement for 498 texts

We also examined that:

1. it is hard to detect manually if a text is fake or real without additional information - facts and context that human annotators may be aware/not aware of (examples: news about domestic policy in different countries). Therefore annotations might become subjective and it is harder to estimate the model's results. The claims verification module for Russian should be the next step of the research;

2. satirical texts can be detected manually better from real news without additional information, based

only on their text: it contains absurd (examples: news about an alien in the Zoo of Perm city (Russia), news about a bank payment terminal on the International Space Station);

3. among 498 annotated texts, most texts about statistics and economics data or accidents are real (examples: news about currency exchange markets and oil prices, news about car accidents statistics). Only one such text is fake. This peculiarity may differ in the real-world data;

4. texts in three classes can be biased: they may contain loaded language, opinion pieces, biased quotations (examples: news about politics in the Middle East, news about Russia-Ukraine conflict). Further research might be focused on hyperpartisan and satirical news and biased language detection;

5. the datasets used in the study should be double-checked, to be unbiased. It concerns mostly texts with questionable quotations and texts with small fragments of fake content. More proper annotation guidelines should be developed, i.e. to handle such cases: the quotation is correct, but it is not truthful;

6. among 498 annotated texts, there were no satirical texts about military news, so deceptive texts could be only fake. The datasets should contain various topics and be taken from different sources, to avoid overfitting.

7   Conclusions

The research can be considered as the first step in building an automated state-of-the-art system for Russian that could detect fake and satirical news and perform automated fact-checking. In this step, we studied language features for fake news and satire detection. We trained and compared different models for fake news detection in Russian, based on all existing available datasets for the Russian language. We investigated bag-of-n-grams features, bag of RST features, and BERT embeddings. We found out that satirical news should be singled out as a separate class, among fake news and real news. We also compared the score of our models with the human score.

The best BERT-based model achieved a 82.2% F1-score and 74.8% accuracy score on a 3 class classification task, which is bigger than the mean human result, but less than the metrics for the bag-of-n-grams based model, which achieved 90.8% F-score and 94.1% accuracy. We found that "bag-of-rst" features do not improve the performance of the bag-of-n-grams model, as they have reached almost the same scores on the test set. Further feature importance analysis and hyperparameter results analysis showed that unigrams are the most important features for fake news detection on our dataset. The model outperforms human evaluation results based on the majority vote.

For automated fake news detection, a combination of different methods should be applied. In future studies, the claims verification module for Russian should be developed and used together with the linguistic features models. New social media datasets of fake, satirical, biased, and hyperpartisan news for Russian should be collected and annotated according to detailed guidelines. We are also planning to try self-supervision methods for extending the datasets. After that, experiments should be performed on creating models for biased and hyperpartisan content, satirical content detection. Wider sets of language and content features should be used for them. We are also going to use multilingual sentence embeddings and transfer learning techniques, in order to incorporate the existing models and approaches, developed for automated claims verification for English, to this task for Russian.

Acknowledgements

The study was funded by Russian Foundation for Basic Research according to the research project No 17-29-07033.

The research was carried out using infrastructure of shared research facilities CKP "Computer science" of FRCCSC RAS. Regulations of CKP "Computer science" // Available at: http://www.frccsc.ru/ckp (date of the application 01.11.2020).

References

O. Ajao, D. Bhowmik, and S. Zargari. 2019. Sentiment aware fake news detection on online social networks. In *ICASSP 2019 - 2019 IEEE International Conference on Acoustics, Speech and Signal Processing (ICASSP)*, pages 2507–2511.

Pepa Atanasova, Preslav Nakov, Lluís Màrquez, Alberto Barrón-Cedeño, Georgi Karadzhov, Tsvetomila Mihaylova, Mitra Mohtarami, and James Glass. 2019. Automatic fact-checking using context and discourse information. *ACM Journal of Data and Information Quality*, 11(3):12:1–12:27.

Isabelle Augenstein, Christina Lioma, Dongsheng Wang, Lucas Chaves Lima, Casper Hansen, Christian Hansen, and Jakob Grue Simonsen. 2019. MultiFC: A real-world multi-domain dataset for evidence-based fact checking of claims. In *Proceedings of the 2019 Conference on Empirical Methods in Natural Language Processing and the 9th International Joint Conference on Natural Language Processing (EMNLP-IJCNLP)*, pages 4685–4697, Hong Kong, China.

Ramy Baly, Georgi Karadzhov, Dimitar Alexandrov, James Glass, and Preslav Nakov. 2018. Predicting factuality of reporting and bias of news media sources. In *Proceedings of the 2018 Conference on Empirical Methods in Natural Language Processing*, pages 3528–3539, Brussels, Belgium.

Mikhail Burtsev, Alexander Seliverstov, Rafael Airapetyan, Mikhail Arkhipov, Dilyara Baymurzina, Nickolay Bushkov, Olga Gureenkova, Taras Khakhulin, Yuri Kuratov, Denis Kuznetsov, Alexey Litinsky, Varvara Logacheva, Alexey Lymar, Valentin Malykh, Maxim Petrov, Vadim Polulyakh, Leonid Pugachev, Alexey Sorokin, Maria Vikhreva, and Marat Zaynutdinov. 2018. DeepPavlov: Open-source library for dialogue systems. In *Proceedings of ACL 2018, System Demonstrations*, pages 122–127, Melbourne, Australia.

Giovanni Da San Martino, Seunghak Yu, Alberto Barrón-Cedeño, Rostislav Petrov, and Preslav Nakov. 2019. Fine-grained analysis of propaganda in news article. In *Proceedings of the 2019 Conference on Empirical Methods in Natural Language Processing and the 9th International Joint Conference on Natural Language Processing (EMNLP-IJCNLP)*, pages 5636–5646, Hong Kong, China.

Sohan De Sarkar, Fan Yang, and Arjun Mukherjee. 2018. Attending sentences to detect satirical fake news. In *Proceedings of the 27th International Conference on Computational Linguistics*, pages 3371–3380, Santa Fe, New Mexico, USA.

Leon Derczynski, Kalina Bontcheva, Maria Liakata, Rob Procter, Geraldine Wong Sak Hoi, and Arkaitz Zubiaga. 2017. SemEval-2017 task 8: RumourEval: Determining rumour veracity and support for rumours. In *Proceedings of the 11th International Workshop on Semantic Evaluation (SemEval-2017)*, pages 69–76, Vancouver, Canada.

Jacob Devlin, Ming-Wei Chang, Kenton Lee, and Kristina Toutanova. 2019. BERT: Pre-training of deep bidirectional transformers for language understanding. In *Proceedings of the 2019 Conference of the North American Chapter of the Association for Computational Linguistics: Human Language Technologies, Volume 1 (Long and Short Papers)*, pages 4171–4186, Minneapolis, Minnesota.

Sebastian Dungs, Ahmet Aker, Norbert Fuhr, and Kalina Bontcheva. 2018. Can rumour stance alone predict veracity? In *Proceedings of the 27th International Conference on Computational Linguistics*, pages 3360–3370, Santa Fe, New Mexico, USA.

Andreas Hanselowski, Hao Zhang, Zile Li, Daniil Sorokin, Benjamin Schiller, Claudia Schulz, and Iryna Gurevych. 2018. UKP-athene: Multi-sentence textual entailment for claim verification. In *Proceedings of the First Workshop on Fact Extraction and VERification (FEVER)*, pages 103–108, Brussels, Belgium.

Georgi Karadzhov, Pepa Gencheva, Preslav Nakov, and Ivan Koychev. 2017. We built a fake news / click bait filter: What happened next will blow your mind! In *Proceedings of the International Conference Recent Advances in Natural Language Processing, RANLP 2017*, pages 334–343, Varna, Bulgaria.

Hamid Karimi and Jiliang Tang. 2019. Learning hierarchical discourse-level structure for fake news detection. In *Proceedings of the 2019 Conference of the North American Chapter of the Association for Computational Linguistics: Human Language Technologies, Volume 1 (Long and Short Papers)*, pages 3432–3442, Minneapolis, Minnesota.

Dhruv Khattar, Jaipal Singh Goud, Manish Gupta, and Vasudeva Varma. 2019. Mvae: Multimodal variational autoencoder for fake news detection. In *WWW '19: The World Wide Web Conference*, pages 2915–2921, San Francisco, CA, USA.

Elena Kochkina and Maria Liakata. 2020. Estimating predictive uncertainty for rumour verification models. In *Proceedings of the 58th Annual Meeting of the Association for Computational Linguistics*, pages 6964–6981, Online.

Elena Kochkina, Maria Liakata, and Arkaitz Zubiaga. 2018. All-in-one: Multi-task learning for rumour verification. In *Proceedings of the 27th International Conference on Computational Linguistics*, pages 3402–3413, Santa Fe, New Mexico, USA, August.

Yuri Kuratov and Mikhail Arkhipov. 2019. Adaptation of deep bidirectional multilingual transformers for russian language. *arXiv preprint arXiv:1905.07213*.

Nayeon Lee, Belinda Li, Sinong Wang, Wen-tau Yih, Hao Ma, and Madian Khabsa. 2020. Language models as fact checkers? In *Proceedings of the Third Workshop on Fact Extraction and VERification (FEVER)*, pages 36–41, Online.

Or Levi, Pedram Hosseini, Mona Diab, and David Broniatowski. 2019. Identifying nuances in fake news vs. satire: Using semantic and linguistic cues. In *Proceedings of the Second Workshop on Natural Language Processing for Internet Freedom: Censorship, Disinformation, and Propaganda*, pages 31–35, Hong Kong, China.

Lisha Li, Kevin Jamieson, Giulia DeSalvo, Afshin Rostamizadeh, and Ameet Talwalkar. 2016. Hyperband: A novel bandit-based approach to hyperparameter optimization.

Jing Ma, Wei Gao, and Kam-Fai Wong. 2018. Rumor detection on twitter with tree-structured recursive neural networks. In *Proceedings of the 56th Annual Meeting of the Association for Computational Linguistics (Volume 1: Long Papers)*, pages 1980–1989, Melbourne, Australia.

Jing Ma, Wei Gao, Shafiq Joty, and Kam-Fai Wong. 2019. Sentence-level evidence embedding for claim verification with hierarchical attention networks. In *Proceedings of the 57th Annual Meeting of the Association for Computational Linguistics*, pages 2561–2571, Florence, Italy.

William C Mann and Sandra A Thompson. 1988. Rhetorical structure theory: Toward a functional theory of text organization. *Text-interdisciplinary Journal for the Study of Discourse*, 8(3):243–281.

Yixin Nie, Haonan Chen, and Mohit Bansal. 2019. Combining fact extraction and verification with neural semantic matching networks. In *Proceedings of the AAAI Conference on Artificial Intelligence*, pages 6859–6866.

Verónica Pérez-Rosas, Bennett Kleinberg, Alexandra Lefevre, and Rada Mihalcea. 2018. Automatic detection of fake news. In *Proceedings of the 27th International Conference on Computational Linguistics*, pages 3391–3401, Santa Fe, New Mexico, USA.

Dina Pisarevskaya. 2017. Deception detection in news reports in the Russian language: Lexics and discourse. In *Proceedings of the 2017 EMNLP Workshop: Natural Language Processing meets Journalism*, pages 74–79, Copenhagen, Denmark.

Beatrice Portelli, Jason Zhao, Tal Schuster, Giuseppe Serra, and Enrico Santus. 2020. Distilling the evidence to augment fact verification models. In *Proceedings of the Third Workshop on Fact Extraction and VERification (FEVER)*, pages 47–51, Online.

Martin Potthast, Johannes Kiesel, Kevin Reinartz, Janek Bevendorff, and Benno Stein. 2018. A stylometric inquiry into hyperpartisan and fake news. In *Proceedings of the 56th Annual Meeting of the Association for Computational Linguistics (Volume 1: Long Papers)*, pages 231–240, Melbourne, Australia.

Hannah Rashkin, Eunsol Choi, Jin Yea Jang, Svitlana Volkova, and Yejin Choi. 2017. Truth of varying shades: Analyzing language in fake news and political fact-checking. In *Proceedings of the 2017 Conference on Empirical Methods in Natural Language Processing*, pages 2931–2937, Copenhagen, Denmark.

Marco Tulio Ribeiro, Sameer Singh, and Carlos Guestrin. 2016. " why should i trust you?" explaining the predictions of any classifier. In *Proceedings of the 22nd ACM SIGKDD international conference on knowledge discovery and data mining*, pages 1135–1144.

Victoria L. Rubin and Tatiana Lukoianova. 2015. Truth and deception at the rhetorical structure level. *Journal of the Association for Information Science and Technology*, 66(5):905–917.

Victoria Rubin, Niall Conroy, Yimin Chen, and Sarah Cornwell. 2016. Fake news or truth? using satirical cues to detect potentially misleading news. In *Proceedings of the Second Workshop on Computational Approaches to Deception Detection*, pages 7–17, San Diego, California.

Natali Ruchansky, Sungyong Seo, and Yan Liu. 2017. Csi: A hybrid deep model for fake news detection. In *CIKM '17: Proceedings of the 2017 ACM on Conference on Information and Knowledge Management*, pages 797–806.

Tal Schuster, Roei Schuster, Darsh J. Shah, and Regina Barzilay. 2020. The limitations of stylometry for detecting machine-generated fake news. *Comput. Linguistics*, 46(2):499–510.

Artem Shelmanov, Dina Pisarevskaya, Elena Chistova, Svetlana Toldova, Maria Kobozeva, and Ivan Smirnov. 2019. Towards the data-driven system for rhetorical parsing of Russian texts. In *Proceedings of the Workshop on Discourse Relation Parsing and Treebanking 2019*, pages 82–87, Minneapolis, MN, June.

Jasper Snoek, Hugo Larochelle, and Ryan P Adams. 2012. Practical bayesian optimization of machine learning algorithms. In *Advances in neural information processing systems*, pages 2951–2959.

Amir Soleimani, Christof Monz, and Marcel Worring. 2020. Bert for evidence retrieval and claim verification. In *Advances in Information Retrieval. ECIR 2020. Lecture Notes in Computer Science, vol 12036*, pages 359–366.

Dominik Stammbach, Stalin Varanasi, and Guenter Neumann. 2019. DOMLIN at SemEval-2019 task 8: Automated fact checking exploiting ratings in community question answering forums. In *Proceedings of the 13th International Workshop on Semantic Evaluation*, pages 1149–1154, Minneapolis, Minnesota, USA.

James Thorne, Andreas Vlachos, Christos Christodoulopoulos, and Arpit Mittal. 2018. FEVER: a large-scale dataset for fact extraction and VERification. In *Proceedings of the 2018 Conference of the North American Chapter of the Association for Computational Linguistics: Human Language Technologies, Volume 1 (Long Papers)*, pages 809–819, New Orleans, Louisiana.

Svitlana Volkova, Kyle Shaffer, Jin Yea Jang, and Nathan Hodas. 2017. Separating facts from fiction: Linguistic models to classify suspicious and trusted news posts on twitter. In *Proceedings of the 55th Annual Meeting of the Association for Computational Linguistics (Volume 2: Short Papers)*, pages 647–653, Vancouver, Canada.

William Yang Wang. 2017. "liar, liar pants on fire": A new benchmark dataset for fake news detection. In *Proceedings of the 55th Annual Meeting of the Association for Computational Linguistics (Volume 2: Short Papers)*, pages 422–426, Vancouver, Canada.

Thomas Wolf, Lysandre Debut, Victor Sanh, Julien Chaumond, Clement Delangue, Anthony Moi, Pierric Cistac, Tim Rault, R'emi Louf, Morgan Funtowicz, and Jamie Brew. 2019. Huggingface's transformers: State-of-the-art natural language processing. *ArXiv*, abs/1910.03771.

Lianwei Wu, Yuan Rao, Haolin Jin, Ambreen Nazir, and Ling Sun. 2019. Different absorption from the same sharing: Sifted multi-task learning for fake news detection. In *Proceedings of the 2019 Conference on Empirical Methods in Natural Language Processing and the 9th International Joint Conference on Natural Language Processing (EMNLP-IJCNLP)*, pages 4644–4653, Hong Kong, China.

Lianwei Wu, Yuan Rao, yongqiang zhao, Hao Liang, and Ambreen Nazir. 2020. DTCA: Decision tree-based co-attention networks for explainable claim verification. In *Proceedings of the 58th Annual Meeting of the Association for Computational Linguistics*, pages 1024–1035, Online.

Fan Yang, Arjun Mukherjee, and Eduard Dragut. 2017. Satirical news detection and analysis using attention mechanism and linguistic features. In *Proceedings of the 2017 Conference on Empirical Methods in Natural Language Processing*, pages 1979–1989, Copenhagen, Denmark.

Alsu Zaynutdinova, Dina Pisarevskaya, Maxim Zubov, and Ilya Makarov. 2019. Deception detection in online media. In *Proceedings of the Fifth Workshop on Experimental Economics and Machine Learning at the National Research University Higher School of Economics co-located with the Seventh International Conference on Applied Research in Economics (iCare7)*, pages 121–127.

Wanjun Zhong, Jingjing Xu, Duyu Tang, Zenan Xu, Nan Duan, Ming Zhou, Jiahai Wang, and Jian Yin. 2020. Reasoning over semantic-level graph for fact checking. In *Proceedings of the 58th Annual Meeting of the Association for Computational Linguistics*, pages 6170–6180, Online.

Arkaitz Zubiaga, Maria Liakata, Rob Procter, Geraldine Wong Sak Hoi, and Peter Tolmie. 2016. Analysing how people orient to and spread rumours in social media by looking at conversational threads. *PLoS ONE11(3)*, pages 1–29.

Appendix A. List of hyperparameters.

Hyperparameter	Distribution/Range
C	uniform(0.001, 10)
gamma	uniform(0.0001, 1)
minimal n-gram frequency	uniform(1, 10)
n-gram's range	uniform, from 1..5-gram to 1..20-gram)
tokenization	with or without lemmatization
k	uniform(100, 10000)
max number of iterations	uniform(100, 10000)
penalty(for logreg)	l1 and elasticnet

Table 8: Hyperparameter space for SVM based classifier with RBF kernel and Logistic Regression based classifier

Model	N-gram range	topK	min frequency	C	γ	penalty
SVM - binary - baseline	(1, 3)	1048	1	2.871	0.8217	-
SVM - binary - RST	(1, 2)	5830	1	9.294	0.321	-
LogReg - binary - baseline	(1, 2)	2640	1	9.975	-	l1
LogReg - binary - RST	(1, 2)	2760	1	9.415	-	elasticnet
SVM - 3 class - baseline	(1,2)	9745	10	9.126	0.4708	-
SVM - 3 class - rst	(1, 15)	8194	8	9.940	0.1729	-
LogReg - 3 class - baseline	(1, 5)	8671	8	9.681	-	elasticnet
LogReg - 3 class - rst	(1, 4)	8411	9	9.781	-	l1

Table 9: Hyperparameter values

Automatic Detection of Hungarian Clickbait and Entertaining Fake News

Veronika Vincze[1] and **Martina Katalin Szabó**[2,3]
[1]MTA-SZTE Research Group on Artificial Intelligence, Szeged, Hungary
[2]Computational Social Science Research Center for Educational and Network Studies
(CSS-RECENS), Centre for Social Sciences, Budapest, Hungary
[3]Department of Software Engineering, University of Szeged, Szeged, Hungary
{vinczev, martina}@inf.u-szeged.hu

Abstract

Online news do not always come from reliable sources and they are not always even realistic. The constantly growing number of online textual data has raised the need for detecting deception and bias in texts from different domains recently. In this paper, we identify different types of unrealistic news (clickbait and fake news written for entertainment purposes) written in Hungarian on the basis of a rich feature set and with the help of machine learning methods. Our tool achieves competitive scores: it is able to classify clickbait, fake news written for entertainment purposes and real news with an accuracy of over 80%. It is also highlighted that morphological features perform the best in this classification task.

1 Introduction

The growing number of online news and the ability to easily and rapidly distribute information on the internet increasingly stimulate demand for automatic fact checking (Thorne and Vlachos, 2018). As a consequence, linguistic aspects of deception, bias and uncertainty detection have raised worldwide interest recently and have been thoroughly studied in a variety of NLP-applications (Zhou et al., 2004; Mihalcea and Strapparava, 2009; Szarvas et al., 2012; Choi et al., 2012; Girlea et al., 2016; Barrón-Cedeño et al., 2019). However, determining the trustworthiness of news and separating facts from misinformation is still a challenging and often controversial task (Graves, 2018).

There may be several motivations for spreading fake news on the internet. Hoax websites, for instance, publish dubious news (clickbait) in order to spread different kinds of misinformation or make money by spreading commercials. To be more specific, fabricated news draw disproportionate attention on social networks most of the time, outperforming conventional news (Graves, 2018). Consequently, publishing interesting fake news is a great way to spread advertisements on the internet. At the same time, there are pages where the primary purpose is to entertain readers by spreading fake news. In this case the readers are aware of the deceptive nature of the information provided and they read it just for fun.

In this paper we will focus on two types of fake news written in Hungarian, namely clickbait and fake news written for entertainment purposes. Our main goal is to distinguish these two types from real news with machine learning methods. We will also analyse what type of linguistic information can contribute the most to performance.

The paper is structured as follows: First, we will give a short review of the relevant research work and we detail the importance and benefits of the recent analysis. Then, we will present the corpus analysed, along with its basic statistical data and the methods and tools we used in the experiments. Next, we will introduce our rich feature set consisting of statistical, morphological, syntactic, semantic and pragmatic features applied for statistical significance analysis and machine learning experiments. We will discuss the findings of the significance analysis, and then, we will provide a detailed description of the machine learning experiments, along with the results. Finally, we discuss the results and add some ideas for future work.

Proceedings of the 3rd International Workshop on Rumours and Deception in Social Media (RDSM), pages 58–69
Barcelona, Spain (Online), December 12, 2020.

2   Related work

Most of the authors address the issue of automatic fact extraction and binary classification verification task based on machine learning methods and annotated datasets. There are several studies addressing the phenomena of deception and bias in different types of discourse. For instance, most of the authors analyze the phenomena of deception and bias either in speech text (Fetzer, 2008; Fraser, 2010; Scheithauer, 2007; Simon-Vandenbergen et al., 2007) or address the issue of automatic fact extraction and binary classification verification task (Greene and Resnik, 2009; Rubin et al., 2015; Wang, 2017; Thorne and Vlachos, 2018; Thorne et al., 2018; Graves, 2018). In addition, uncertainty detectors have mostly been developed in the biological and medical domains (Szarvas et al., 2012). Also, a few studies seem to address the issue of the systematic analysis of a huge amount of propaganda texts (Propaganda Analysis, 1938; Rashkin et al., 2017; Barrón-Cedeño et al., 2019; Vincze et al., 2019; Kmetty et al., 2020; Szabó et al., 2020).

Recently, several databases have been compiled for the English language which contain both fake and real news. For instance, the dataset described in Vlachos and Riedel (2014) consists of a total of 221 statements, along with their veracity value. The Liar corpus contains approximately 13,000 short political statements (Wang, 2017). The Emergent Corpus consists of 300 statements and 2500 news related to the semantic content of the statements, along with their veracity value (Ferreira and Vlachos, 2016). Our study is most similar in vein to Rubin et al. (2016), which compares satirical news to real news, collected from different websites. Pérez-Rosas et al. (2018) used crowdsourcing to generate fake news on the basis of real news and then carried out machine learning experiments to separate the two types. There are also a few studies that identify clickbaits (see e.g. Biyani et al. (2016) for English and Karadzhov et al. (2017) for Bulgarian).

The above studies focus on the phenomena of deception and bias almost exclusively in sources written in English, so the reported research findings and models are mostly limited to the English language. The same goes for the currently existing datasets. Santos et al. (2020) forms an exception, which, however, aims at the distinction of Brazilian Portuguese fake and real news. To the best of our knowledge, our recent research work is the first attempt at the automatic detection of Hungarian fake news.

The main contributions of our paper are the following:

- We present a novel dataset for detecting fake news in Hungarian;

- We define a rich feature set of linguistic parameters for detecting different characteristics of different types of Hungarian news examined;

- We carry out a detailed statistical analysis of linguistic parameters that may distinguish real news, clickbait and entertaining fake news in Hungarian;

- We perform machine learning experiments with the above mentioned feature set for detecting real news, clickbait and fake news written for entertainment purposes;

- We analyze the effect of each set of features and identify the best combination of these features for the task.

3   The corpus

Our study was conducted on a corpus of 180 online news compiled by us. First, the real news come from several national and regional news portals, e.g. `www.index.hu`, `www.origo.hu` and `www.delmagyar.hu`. Second, the fake news were collected from two sources. On the one hand, we selected some special news published on the 1st of April (Fools' Day) on some of the news portals basically spreading real news, as this day is traditionally seen as an occasion for making pranks. These news were intended to play a trick on readers. On the other hand, we downloaded articles from the Hírcsárda website[1]. These texts are mostly based on parodies of current political and public events, and their

---

[1] `https://www.hircsarda.hu`

Texts	Number of texts	Number of sentences	Number of tokens
Real news	60	1673	22203
Clickbait	60	1241	13925
Entertaining fake news	60	1310	16193

Table 1: Basic statistical data of the corpus.

explicit purpose is not to spread real or fake news but to entertain the readers. Third, the clickbait news were downloaded from websites that were collected by a Hungarian real news portal and were claimed to be unreliable[2]. Most of the texts were published during the time period between 2017 and 2019. For each type of text, we collected 60 documents. The news were randomly selected but we made sure that a given topic should not be included in the corpus more than one time.

The basic data of our corpus are presented in Table 1.

4 Experiments

In this section, we present our methods for identifying the three types of news with machine learning methods as well as providing statistical significance analysis for the linguistic features defined for the task.

4.1 Feature set

As a first step, we automatically preprocessed the texts with magyarlanc (Zsibrita et al., 2013), a toolkit written in JAVA for the linguistic processing of Hungarian texts. With this tool, the text was first split into sentences, then tokenised and lemmatised, and finally morphologically and syntactically analyzed. Based on the output of magyarlanc, we extracted a high number of linguistic features. Our feature set consists of the following features:

- **Basic statistical features**: the number of sentences; the number of words; the number and rate of lemmas; the average sentence length.

- **Morphological features**: Part-of-speech (or POS) features: the number and frequency of nouns, proper nouns, verbs, adjectives, (demonstrative) pronouns, numerals, adverbs, conjunctions and unanalyzed words (i.e. those with an "unknown" POS tag); the number of punctuation marks; the number and frequency of imperative and conditional verbs; the number and frequency of past and present tense verbs; the number and frequency of first person singular verbs and of first person plural verbs; the number and frequency of frequentative, causative and modal verbs; the number and frequency of comparative and superlative adjectives.

- **Syntactic features**: the number and frequency of subjects and objects as Hungarian is a pro-drop language, meaning that pronominal subjects and objects might not be overt in the clause; the number and frequency of adverbials; the number and frequency of coordinations and subordinations.

- **Semantic features**: the number and frequency of negation words; the number and frequency of content words and function words; the number of frequency of public verbs, private verbs and suasive verbs based on a Hungarian translation of lists found in Quirk et al. (1985). Uncertainty features: the number and frequency of words belonging to several classes of linguistic uncertainty based on Vincze (2014a). Sentiment features: the number and frequency of positive and negative words based on a list of sentiment phrases. We applied two different Hungarian dictionaries for sentiment analysis: one list was a translation of Liu (2012), while the other one contained Hungarian slang words (Szabó, 2015), respectively. Emotion features: the number and frequency of words belonging to the emotions described in Szabó et al. (2016).

  For list-based semantic features we used a simple dictionary-based method. Thus, if a lemma in our corpora matched with any item in our lists, it was counted as a hit.

---

[2] https://hvg.hu/tudomany/20150119_atveros_weboldalak

- **Pragmatic features**: the number and frequency of speech act verbs, based on a list manually constructed by us; the number and frequency of literal quotes and citations, detected on the basis of quotation marks and dashes at the beginning of the sentences. The number and frequency of discourse markers. To find discourse markers in the texts we applied a word list based on Dér and Markó (2007).

4.2 Statistical significance analysis

In order to measure the effect of each feature in distinguishing real news, clickbait and entertaining fake news, we performed statistical significance tests (pairwise t-tests) for all features.

Here we assume that the distinction between different types of fake news and real news is primarily a semantic-pragmatic problem. As a consequence, we also presuppose that morphological and syntactic features of the texts in the three subcorpora will not necessarily differ from each other.

The results of the statistical significance analysis are shown in Table 2. For the sake of simplicity, we provide p-values for features only where a statistically significant difference was found. For comparison, we also report the mean values for each feature for all classes in Table 3.

4.3 Machine learning experiments

In addition to the statistical significance analysis, we seek to automatically discriminate clickbait and fake news from real news. For this purpose, we made use of the above mentioned rich feature set including statistical, morphological, syntactic, semantic and pragmatic characteristics of Hungarian texts.

In our experiments, we used a Random Forest classifier (Breiman, 2001) with ten fold cross validation since it does not easily overfit. . In order to examine which features play the most important role in distinguishing the three groups of news, we divided the features into five main groups based on a linguistic classification, and experimented with all possible combinations of these groups, yielding an extensive ablation analysis. The baseline method was majority classification, which achieved an accuracy of 33.33%.

5 Results

In this section, we present the results of both our statistical analysis and machine learning experiments.

5.1 Results of statistical significance analysis

Table 2 shows the features that exhibit significant differences among three groups of news. Basically, our hypothesis is just partly confirmed: the findings do not show the outstanding role of semantic-pragmatic features because at the same time, morphological characteristics of the tokens proved to be also essential.

Apart from morphological features, the frequency distribution of certain types of semantic contents also seem to be significantly different in the three subcorpora. For instance, there is a significant difference of the frequency of uncertainty markers such as condition, weasel, peacock and hedge between clickbait news and both of the other corpora. While the first type belongs to semantic uncertainty, the latter three are types of uncertainty at the discourse level (Vincze, 2013). In the case of semantic uncertainty, the lexical content (meaning) of the uncertainty marker (cue) is responsible for uncertainty, e.g. *may*, *possible*, *believe* etc. (Vincze, 2014b). In contrast to semantic uncertainty (Szarvas et al., 2012), in the case of discourse level uncertainty, "the missing or intentionally omitted information is not related to the propositional content of the utterance but to other factors", e.g. for cues like *some*, *often*, *much* etc. Bias evoked by discourse-level uncertainty might be viewed as a characteristic feature of clickbait news.

5.2 Results of machine learning experiments

In the best scenario, our machine learning algorithm achieved an accuracy of 82.78%, which was yielded by combining statistical, morphological, semantic and pragmatic features. This proved to be the best combination of features for identifying fake news (with an F-score of 81.4). The combination of morphological and semantic features seemed to be the most effective for identifying real news, obtaining an F-score of 80.3. As for clickbait news, the combination of statistical, syntactic, semantic and pragmatic features yielded the best result (an F-score of 89.1). More detailed results are shown in Table 4.

Feature	click-fake	fake-real	click-real	Feature	click-fake	fake-real	click-real
token #			0.0047	coord. %			0.0306
lemma #	0.0054		0.0013	uncertain %	0.0254		0.0053
lemma %	<0.0001	0.0001		negation %	0.0003		<0.0001
token %			0.0013	invest. #			0.0054
sentence length			0.0007	epistemic %	0.0492		
unknown #	<0.0001	<0.0001		invest. %			0.0094
unknown %	<0.0001	<0.0001		condition %	0.0154		0.0005
noun #	0.0328	0.0379	0.0009	weasel %	0.0178		0.0067
adjective #	0.0137	0.0456	0.0008	peacock %	0.0088		0.0040
pronoun #	0.0213			hedge %	0.0062		<0.0001
numeral #		0.0006	0.0001	joy #	0.0083		0.0102
punct			0.0149	sorrow #	0.0045	0.0253	
proper noun #	<0.0001		<0.0001	love #	0.0159		0.0248
verb %	<0.0001	0.0358	<0.0001	joy %	<0.0001	0.0011	<0.0001
noun %	<0.0001		<0.0001	fear %	0.0310		0.0021
adjective %	0.0001		<0.0001	sorrow %	0.0058	0.0117	
pronoun %	<0.0001		<0.0001	love %	0.0078		0.0007
conjunction %	0.0094		0.0012	anxiety %	0.0288		0.0063
numeral %		0.0005	<0.0001	surprise %		0.0497	0.0030
adverb %	<0.0001	0.0007	<0.0001	positive1 #	0.0273		
proper noun %	<0.0001		<0.0001	negative1 #		0.0221	
comparative #			0.0326	positive2 #		0.0279	0.0280
superlative %	0.0214			positive2 %	<0.0001		<0.0001
past tense #		0.0154		negative1 %	0.0010		
past tense %		0.0018	0.0165	positive2 %	0.0001	0.0011	<0.0001
present tense %		0.0035		emo. negative %	0.0280		0.0125
imperative #	0.0009		0.0286	content %	<0.0001		<0.0001
imperative %	0.0006		0.0008	function %	<0.0001		<0.0001
1Sg verb %		0.0046	0.0255	public verb #		0.0047	0.0002
dem. pronoun #	0.0093	0.0026		private verb %			0.0050
1Pl verb %		0.0357		public verb %		0.0228	0.0003
dem. pronoun %		0.0067	0.0080	suasive verb %	0.0245		
modal verb %		0.0019	0.0440	quote #	<0.0001		0.0206
subject #		0.0418	0.0071	dash #	0.0011		0.0178
subord. #	0.0056	0.0213	0.0002	speech act %	0.0322		0.0179
coord. #			0.0090	quote %	<0.0001	0.0008	
subord. %	<0.0001	0.0005	<0.0001	dash %	0.0058		
adverbial %	0.0259		0.0246	disc. marker %	0.0001	0.0034	<0.0001

Table 2: Statistically significant features. #: number, %: rate.

Feature	click	fake	real	Feature	click	fake	real
token #	232.0833	269.8833	370.0500	uncertain #	2.8000	2.2167	2.8667
sentence #	20.6833	21.8333	27.8833	uncertain %	0.0119	0.0081	0.0073
lemma #	143.0167	188.0333	211.9667	negation #	4.4333	3.6167	4.3500
lemma %	0.6441	0.7334	0.6595	negation %	0.0199	0.0125	0.0104
token %	13.9728	14.8909	15.8212	epistemic #	0.7500	0.6000	0.8333
sentence length	11.6967	12.5064	13.3731	investigation #	0.0500	0.2000	0.4167
unknown #	0.0500	1.3000	0.2333	condition #	2.2333	1.5167	1.9833
unknown %	0.0003	0.0052	0.0005	weasel #	7.2667	7.0000	7.4667
verb #	38.2333	34.9500	44.9167	peacock #	1.8000	1.2000	1.6500
noun #	58.0500	75.7167	107.4667	hedge #	3.6333	3.2667	3.8167
adjective #	24.6000	35.3333	52.7167	doxastic #	2.8000	2.1333	3.2667
pronoun #	17.5667	12.7667	18.0833	epistemic %	0.0035	0.0019	0.0019
conjunction #	6.9333	6.2333	8.3500	investigation %	0.0002	0.0007	0.0016
numeral #	6.4167	7.4000	15.0833	condition %	0.0096	0.0059	0.0044
adverb #	25.6000	24.7667	29.5667	weasel %	0.0311	0.0246	0.0236
punct #	46.1500	53.1500	69.8667	peacock %	0.0077	0.0045	0.0044
proper noun#	4.8667	17.8167	20.9500	hedge %	0.0157	0.0112	0.0091
verb %	0.1682	0.1299	0.1195	doxastic %	0.0117	0.0086	0.0094
noun %	0.2451	0.2784	0.2879	joy #	4.1000	2.2833	2.1333
adjective %	0.1034	0.1269	0.1347	fear #	0.7000	0.4667	0.3833
pronoun %	0.0748	0.0454	0.0465	anger #	0.4000	0.2333	0.5333
conjunction %	0.0304	0.0232	0.0216	sorrow #	1.1000	0.3167	3.0500
numeral %	0.0269	0.0298	0.0467	love #	0.8500	0.3833	0.4000
adverb %	0.1165	0.0888	0.0740	anxiety #	0.5500	0.3333	0.3500
proper noun %	0.0204	0.0706	0.0661	disgust #	0.1833	0.1833	0.2167
superlative #	0.8333	0.5500	0.6667	surprise #	0.3833	0.2500	0.2167
comparative #	0.7167	0.9167	1.3167	joy %	0.0159	0.0076	0.0041
superlative %	0.0316	0.0138	0.0167	fear %	0.0034	0.0017	0.0010
comparative %	0.0328	0.0271	0.0313	anger %	0.0015	0.0007	0.0017
Sg1 verb #	1.4833	1.1500	1.2167	sorrow %	0.0049	0.0014	0.0050
Pl1 verb #	2.6500	2.5000	2.8333	love %	0.0039	0.0014	0.0008
past #	13.8333	11.9167	19.5500	anxiety %	0.0022	0.0009	0.0007
present #	20.1167	19.6333	21.5167	disgust %	0.0007	0.0005	0.0006
past %	0.3758	0.3546	0.4827	surprise %	0.0019	0.0009	0.0003
present %	0.5167	0.5471	0.4364	positive #	11.9167	8.5000	11.7667
imperative #	3.5500	1.5333	2.0667	negative #	8.4167	6.9000	12.4500
cond. verb #	1.1333	1.7333	1.4833	positive2 #	7.4333	7.2833	13.1167
imperative %	0.0890	0.0396	0.0393	negative2 #	16.5833	14.3000	16.5167
cond. verb %	0.0310	0.0506	0.0312	neg. emotive #	0.4667	0.3333	0.6000
Sg1 verb %	0.0317	0.0310	0.0127	positive %	0.0498	0.0303	0.0275
dem. pronoun #	6.4500	4.3000	7.8500	negative %	0.0370	0.0249	0.0303
Pl1 verb %	0.0736	0.0684	0.0401	positive2 %	0.0700	0.0516	0.0395
dem. pron %	0.3606	0.3533	0.4523	negative2 %	0.0328	0.0269	0.0327
noun morph #	0.8986	0.9023	0.9259	neg.emotive %	0.0027	0.0011	0.0008
freq. verb #	0.0833	0.0667	0.0333	content %	0.6806	0.7244	0.7290
modal verb #	1.9667	2.0667	1.4500	function %	0.2566	0.2145	0.2224
caus. verb #	0.2333	0.1333	0.2833	private verb #	4.4000	3.4833	4.1167
freq. verb %	0.0026	0.0017	0.0010	public verb #	1.4500	1.7167	2.8833
modal verb %	0.0546	0.0602	0.0343	suasive verb #	0.6667	0.8333	1.0333
caus. verb %	0.0074	0.0045	0.0070	private verb %	0.1132	0.0941	0.0822
subject #	17.6500	19.9833	28.2000	public verb %	0.0401	0.0524	0.0741
object #	14.1667	14.6333	16.8833	suasive verb %	0.0145	0.0264	0.0233
attributive #	49.3167	70.3333	105.8333	speech act #	4.2333	3.7667	4.9667
adverbial #	15.7167	15.2167	19.6333	quote #	1.9333	5.8667	4.8500
coordination #	16.3667	18.6000	25.9833	dash #	0.0333	0.5333	0.3000
subject %	0.9167	0.9507	0.9927	speech act %	0.0188	0.0147	0.0143
object %	0.7134	0.6927	0.6702	quote %	0.0084	0.0214	0.0109
attributive %	2.4684	3.2076	3.7817	dash %	0.0002	0.0023	0.0008
adverbial %	0.8044	0.6866	0.6747	discourse marker #	10.4833	9.4500	10.7833
coordination %	0.7947	0.8435	0.9329	discourse marker %	0.0456	0.0338	0.0256

Table 3: Mean values for features in each class. #: number, %: rate.

Feature groups	Acc	clickbait			fake news			real news			all		
		P	R	F	P	R	F	P	R	F	P	R	F
stat+morph+synt+sem+prag	79.44	86.9	88.3	87.6	78.2	71.7	74.8	73.4	78.3	75.8	79.5	79.4	79.4
stat+morph+synt+sem	80.56	86.9	88.3	87.6	78.9	75	76.9	75.8	78.3	77	80.5	80.6	80.5
stat+morph+synt+prag	79.44	85	85	85	80.7	76.7	78.6	73	76.7	74.8	79.6	79.4	79.5
stat+morph+sem+prag	**82.78**	88.3	88.3	88.3	**82.8**	**80**	**81.4**	77.4	80	78.7	**82.8**	**82.8**	**82.8**
stat+synt+sem+prag	75	83.8	**95**	**89.1**	67.9	63.3	65.5	71.4	66.7	69	74.4	75	74.5
morph+synt+sem+prag	81.11	85.2	86.7	86	80.4	75	77.6	**77.8**	81.7	79.7	81.1	81.1	81.1
stat+morph+synt	78.33	85	85	85	77.6	75	76.3	72.6	75	73.8	78.4	78.3	78.3
stat+morph+sem	81.11	85.5	88.3	86.9	82.7	71.7	76.8	75.8	83.3	79.4	81.3	81.1	81
stat+synt+sem	72.78	84.6	91.7	88	66	55	60	66.2	71.7	68.8	72.3	72.8	72.3
morph+synt+sem	80	86.7	86.7	86.7	80	73.3	76.5	73.8	80	76.8	80.2	80	80
stat+morph+prag	79.44	83.6	85	84.3	78.6	73.3	75.9	76.2	80	78	76.2	80	78
stat+synt+prag	67.22	77	78.3	77.7	67.9	63.3	65.5	57.1	60	58.5	67.3	67.2	67.2
morph+synt+prag	77.78	82.3	85	83.6	77.2	73.3	75.2	73.8	75	74.4	77.7	77.8	77.7
stat+sem+prag	75.56	84.4	90	87.1	68.5	61.7	64.9	72.6	75	73.8	75.2	75.6	75.3
morph+sem+prag	81.11	**88.1**	86.7	87.4	80	73.3	76.5	75.8	83.3	79.4	81.3	81.1	81.1
synt+sem+prag	71.67	81.5	88.3	84.8	62.5	58.3	60.3	69.5	68.3	68.9	71.2	71.7	71.4
stat+morph	78.89	87.7	83.3	85.5	78.6	73.3	75.9	71.6	80	75.6	79.3	78.9	79
stat+synt	65.56	77	78.3	77.7	60	55	57.4	59.4	63.3	61.3	65.5	65.6	65.5
stat+sem	73.33	83.6	85	84.3	62.1	60	61	73.8	75	74.4	73.1	73.3	73.2
stat+prag	69.44	77.2	73.3	75.2	70.5	71.7	71.1	61.3	63.3	62.3	69.7	69.4	69.5
morph+synt	78.89	82	83.3	82.6	79.7	78.3	79	75	75	75	78.9	78.9	78.9
morph+sem	81.11	84.4	90	87.1	83.7	68.3	75.2	76.1	**85**	**80.3**	81.4	81.1	80.9
morph+prag	78.89	80.6	83.3	82	84.6	73.3	78.6	72.7	80	76.2	79.3	78.9	78.9
synt+sem	72.22	85	85	85	60.9	65	62.9	71.4	66.7	69	72.5	72.2	72.3
synt+prag	66.11	76.3	75	75.6	59.7	61.7	60.7	62.7	61.7	62.2	66.2	66.1	66.2
sem+prag	71.11	80.3	81.7	81	59.3	58.3	58.8	73.3	73.3	73.3	71	71.1	71
stat	63.33	74.6	78.3	76.4	55.9	55	55.5	58.6	56.7	57.6	63.1	63.3	63.2
morph	80.56	86.2	83.3	84	82.1	76.7	79.3	74.2	81.7	77.8	80.9	80.6	80.6
synt	57.78	71.7	71.7	71.7	43.1	41.7	42.4	58.1	60	59	57.6	57.8	57.7
sem	65.56	78.3	78.3	78.3	54.1	55	54.5	64.4	63.3	63.9	65.6	65.6	65.6
prag	58.89	63.8	61.7	62.7	53.4	65	58.6	61.2	50	55	59.5	58.9	58.8

Table 4: Results of machine learning experiments. Acc: accuracy, P: precision, R: recall, F: F-measure. stat: statistical, morph: morphological, synt: syntactic, sem: semantic, prag: pragmatic.

As can be seen, our algorithm is notably more effective than the baseline method: even the least effective combination of features (i.e. syntactic features on their own) obtained an accuracy of 57.78%. The data also show that the system performs best when it comes to the detection of clickbait news: in the best case scenario, the algorithm properly identified 53 texts and only 2 texts were misclassified as fake news and 5 as real news. The performance was a bit weaker in the case of fake and real news. Overall, the method can be considered effective as it classified only 14 unreal news as real news from the 120 clickbait and fake news.

We also examined the efficiency of each feature set in our ablation experiments. As shown in Table 4, morphological features proved to be most effective, but semantic features also contributed to the success of the automatic detection. The results of the significance tests indicated that syntactic and pragmatic features had a somewhat weaker role in distinguishing the classes, which fact was also confirmed by our machine learning experiments.

6   Discussion of the results

Our results showed that our machine learning experiments achieved an accuracy of 82.78% in the classification task of real, clickbait and entertaining fake news. In this case we used all the feature groups with the exception of syntactic features. Here we discuss our results, with special regard to morphological and semantic features. Moreover, we also provide an error analysis.

6.1   The role of morphology

Examining the role of each feature set, our hypothesis is just partly confirmed: the findings do not show the outstanding role of semantic-pragmatic features in our machine learning experiments. Rather, it was morphology that proved to be the most effective.

In addition to the significant differences of frequency distributions of specific morphological features, morphological characteristics of the subcorpora examined played a decisive role in our machine learning experiments. For instance, when we applied morphological features of the subcorpora exclusively, we achieved an F-score of 80.6 that can be considered notably high compared to the best performance of our algorithm (82.8). At the same time, without these features the best performance we achieved was 75.3% (using statistical, semantic and pragmatic features).

In order to further examine the role of morphology, we divided the features into part-of-speech and deeper morphological features and reran the experiments with only parts-of-speech features and deep morphological features, respectively. This analysis showed that as long as the algorithm achieved 78% using part-of-speech features exclusively, the performance was only 58% when we used the deep morphological features, highlighting the outstanding role of POS tags in the task. Therefore, by using only POS-level information (i.e. keeping the feature set very simple), we can achieve an encouraging result for identifying Hungarian fabricated news.

To investigate this further, we analyzed the frequency distribution of parts-of-speech in a meticulous way. There is a significant difference of the frequency distribution of verbs and adverbs among all the subcorpora examined: there are significantly more verbs and adverbs in clickbait news. What is more, there is a significant difference of the noun, adjective and pronoun rates between clickbait and entertaining fake news, as well as clickbait and real news. More specifically, contributors of clickbait news use less nouns and adjectives and more pronouns than contributors of real news and entertaining fake news. From these results we can conclude that the sentence structure of the clickbait news is notably different from that of the real news: by using more verbal elements, clickbait news seem to highlight the events and happenings and pay less attention to the participants of the events (i.e. nominal elements). In other words, clickbait news emphasize what happened and the details behind the act (e.g. actors, objects etc.) are less noteworthy.

As for other morphological features of the subcorpora, authors of clickbait news use more imperative and less conditional verb forms than the other two text types. While the former feature may be considered as a sign of the need for action, the latter might be viewed as a sign of uncertainty. The results also showed that there is a higher frequency of past tense in real news than present tense compared to clickbait and entertaining fake news. In other words, real and fake news appear to concentrate on past events, i.e. they have a descriptive function, whereas clickbait news focus on the "here and now", representing a more "active" and "powerful" discourse, this enticing readers to click on them. It is also worth mentioning that there is a higher occurrence of first person plural verb forms in clickbait and in entertaining fake news than in real news. This feature may be considered as a linguistic strategy for manipulation since in the unreal news, it may evoke a shared feeling of common ground, attempting to deceive the reader (Bártházi, 2008).

6.2 The role of semantics

As for frequency distribution of certain types of semantic contents there is a significantly higher frequency of uncertainty markers such as condition, weasel, peacock and hedge in clickbait news compared to the fake and real news. At the same time, the frequency difference between fake and real news is not significant in this case. The latter feature shows that fake news tend to present information as factual statements, thereby increasing the apparent authenticity and credibility of the content.

With regard to the sentiment and emotional content of the texts, there is a significantly higher frequency of positive sentiment rate in clickbait news. What is more, there is a significantly higher frequency of "love" and "joy" in clickbait news compared to the other two text types. Data also show that these emotions are more frequent in fake news than in real news as well. Based on these results we may conclude that positive attitude characterizes unreliable news more than real news. These results correlate with a previous research finding about the linguistic features of Hungarian communist propaganda texts (Vincze et al., 2019), where it was also proved that propaganda texts bound in positive emotions.

However, we also should mention that there are more words representing anxiety and fear in clickbait news than real and fake news as well. This might reflect the emphatic role of emotions in clickbait news, which is probably related to the general purpose of these texts, i.e. the more emotional a text is, the more probably it will generate clicks.

6.3   Error analysis

Below we present some examples of news that are misclassified by the system.

Real news classified as fake news:

- *Golden State Warriors are reluctant to celebrate their latest title at the White House*[3]

- *180-year old postcard making company wound up*[4]

Fake news classified as real news:

- *Formula-1 in Szeged, Hungary*[5]

- *Fake Grabovoi numbers appeared*[6]

Clickbait news classified as real news:

- *Scary: Nostradamus's prophecies for 2019*[7]

- *10 shocking photos without an explanation*[8]

Analyzing the incorrectly classified documents, it was revealed that short real news were often misclassified as real news tend to be longer than fake news. On the other hand, clickbait news that contained a lower rate of imperatives and/or a higher rate of conditionals were more prone to be classified as real news. Finally, fake news with lots of negative words were often misclassified as real news.

7   Conclusions and future work

In this work, we reported on the automatic discrimination between Hungarian real news, clickbait news and fake news written for entertaining purposes. Our results confirm that it is possible to successfully detect untrustful news based on a rich – morphological, syntactic, semantic and pragmatic – feature set, and especially morphological features play an important role in the process. Besides morphological features, the added value of semantic features was also apparent, while at the same time, syntactic features did not have a notable effect on our results. Our experiments show the potential for exploiting the morphological, more specifically part-of-speech features for fake news detection.

As a next step of the research, on the basis of the findings of the recent analysis, we would like to apply our methods and tools on texts belonging to other domains. We will attempt to further train our algorithm to discriminate texts containing propagandistic features, bias, misinformation or disinformation. Moreover, we would like to compare these results to previous findings concerning the linguistic features of Hungarian Communist propaganda texts (Vincze et al., 2019). Finally, we would like to compare our results obtained on Hungarian texts with those on English corpora (Barrón-Cedeño et al., 2019) and possibly other languages as well.

Acknowledgements

This work was supported by grantTUDFO/47138-1/2019-ITM of the Ministry for Innovation and Technology, Hungary and by the Hungarian Artificial Intelligence National Laboratory.

---

[3]`http://index.hu/sport/kosarlabda/2017/09/24/donald_trump_sertodes_golden_state_warriors_steph_curry_lebron_james_kobe_bryant/`

[4]`http://www.origo.hu/gazdasag/20170927-180-eves-kepeslap-keszito-csaladi-vallalkozas-szunik-meg.html`

[5]`https://szegedpanorama.blogspot.hu/2013/04/forma-1-es-verseny-szegeden.html`

[6]`http://hircsarda.hu/2015/02/19/mar_hamisitjak_a_grabovoj-szamokat/`

[7]`http://eztnezdmeg.com/hatborzongato-nostradamus-szerint-ez-var-rank-2019-ben/`

[8]`http://www.hirvarazs.info/altalanos/tiz-dobbenetes-foto-amire-sose-talaltak-magyarazatot/`

References

Alberto Barrón-Cedeño, Israa Jaradat, Giovanni Da San Martino, and Preslav Nakov. 2019. Proppy: Organizing the news based on their propagandistic content. *Information Processing Management*, 56(5):1849–1864.

Eszter Bártházi. 2008. Manipuláció, valamint manipulációra alkalmas nyelvhasználati eszközök a sajtóreklámokban. *Magyar Nyelv*, 104(4):443–463.

P. Biyani, K. Tsioutsiouliklis, and John Blackmer. 2016. "8 amazing secrets for getting more clicks": Detecting clickbaits in news streams using article informality. In *AAAI*.

Leo Breiman. 2001. Random forests. *Machine Learning*, 45(1):5–32, October.

Eunsol Choi, Chenhao Tan, Lillian Lee, Cristian Danescu-Niculescu-Mizil, and Jennifer Spindel. 2012. Hedge detection as a lens on framing in the gmo debates: A position paper. In *Proceedings of the Workshop on Extra-Propositional Aspects of Meaning in Computational Linguistics*, pages 70–79. Association for Computational Linguistics.

Csilla Ilona Dér and Alexandra Markó. 2007. A magyar diskurzusjelölők szupraszegmentális jelöltsége. In *Nyelvelmélet–nyelvhasználat*, pages 61–67. Tinta, Székesfehérvár–Budapest.

William Ferreira and Andreas Vlachos. 2016. Emergent: a novel data-set for stance classification. In *Proceedings of the 2016 Conference of the North American Chapter of the Association for Computational Linguistics: Human Language Technologies*, pages 1163–1168, San Diego, California, June. Association for Computational Linguistics.

Anita Fetzer. 2008. "And I Think That Is a Very Straightforward Way of Dealing With It"– The Communicative Function of Cognitive Verbs in Political Discourse. *Journal of Language and Social Psychology*, 27:384–396, 12.

Bruce Fraser. 2010. Chapter 11. hedging in political discourse. In *Perspectives in Politics and Discourse*, pages 201–214. 01.

Codruta Girlea, Roxana Girju, and Eyal Amir. 2016. Psycholinguistic features for deceptive role detection in werewolf. In *Proceedings of the 2016 Conference of the North American Chapter of the Association for Computational Linguistics: Human Language Technologies*, pages 417–422.

Lucas Graves. 2018. Understanding the promise and limits of automated fact-checking. Technical report.

Stephan Greene and Philip Resnik. 2009. More than words: Syntactic packaging and implicit sentiment. In *Proceedings of Human Language Technologies: The 2009 Annual Conference of the North American Chapter of the Association for Computational Linguistics*, pages 503–511, Boulder, Colorado, June. Association for Computational Linguistics.

Georgi Karadzhov, Pepa Gencheva, Preslav Nakov, and Ivan Koychev. 2017. We built a fake news / click bait filter: What happened next will blow your mind! *ArXiv*, abs/1803.03786.

Zoltán Kmetty, Veronika Vincze, Dorottya Demszky, Orsolya Ring, Balázs Nagy, and Martina Katalin Szabó. 2020. Pártélet: A hungarian corpus of propaganda texts from the hungarian socialist era. In *Proceedings of The 12th Language Resources and Evaluation Conference*, pages 2381–2388.

Bing Liu. 2012. *Sentiment Analysis and Opinion Mining*. Morgan & Claypool Publishers.

Rada Mihalcea and Carlo Strapparava. 2009. The lie detector: Explorations in the automatic recognition of deceptive language. In *Proceedings of the ACL-IJCNLP 2009 Conference Short Papers*, pages 309–312. Association for Computational Linguistics.

Verónica Pérez-Rosas, Bennett Kleinberg, Alexandra Lefevre, and Rada Mihalcea. 2018. Automatic detection of fake news. In *Proceedings of the 27th International Conference on Computational Linguistics*, pages 3391–3401, Santa Fe, New Mexico, USA, August. Association for Computational Linguistics.

Institute for Propaganda Analysis. 1938. How to Detect Propaganda. In *Propaganda Analysis. Publications of the Institute for Propaganda Analysis*, volume I, pages 210–218.

Randolph Quirk, Sidney Greenbaum, Geoffrey Leech, and Jan Svartvik. 1985. *A Comprehensive Grammar of the English Language*. Longman, London.

Hannah Rashkin, Eunsol Choi, Jin Yea Jang, Svitlana Volkova, and Yejin Choi. 2017. Truth of varying shades: Analyzing language in fake news and political fact-checking. In *Proceedings of the 2017 Conference on Empirical Methods in Natural Language Processing*, pages 2931–2937, Copenhagen, Denmark, September. Association for Computational Linguistics.

Victoria L. Rubin, Yimin Chen, and Niall J. Conroy. 2015. Deception detection for news: Three types of fakes. In *Proceedings of the 78th ASIS&T Annual Meeting: Information Science with Impact: Research in and for the Community*, ASIST '15, pages 83:1–83:4, Silver Springs, MD, USA. American Society for Information Science.

Victoria Rubin, Niall Conroy, Yimin Chen, and Sarah Cornwell. 2016. Fake news or truth? using satirical cues to detect potentially misleading news. In *Proceedings of the Second Workshop on Computational Approaches to Deception Detection*, pages 7–17, San Diego, California, June. Association for Computational Linguistics.

Roney Santos, Gabriela Pedro, Sidney Leal, Oto Vale, Thiago Pardo, Kalina Bontcheva, and Carolina Scarton. 2020. Measuring the impact of readability features in fake news detection. In *Proceedings of the 12th Language Resources and Evaluation Conference*, pages 1404–1413, Marseille, France, May. European Language Resources Association.

Rut Scheithauer. 2007. Metaphors in election night television coverage in Britain, the United States and Germany. In *Political Discourse in the Media: Cross-cultural perspectives*, pages 75–106. 01.

Anne-Marie Simon-Vandenbergen, Peter White, and Karin Aijmer. 2007. Presupposition and 'taking-for-granted' in mass communicated political argument An illustration from British, Flemish and Swedish political colloquy. In *Political Discourse in the Media*, pages 31–74. 01.

Martina Katalin Szabó, Veronika Vincze, and Gergely Morvay. 2016. Magyar nyelvű szövegek emócióelemzésének elméleti nyelvészeti és nyelvtechnológiai problémái. In *Távlatok a mai magyar alkalmazott nyelvészetben*. Tinta, Budapest.

Martina Katalin Szabó, Orsolya Ring, Balázs Nagy, László Kiss, Júlia Koltai, Gábor Berend, László Vidács, Attila Gulyás, and Zoltán Kmetty. 2020. Exploring the dynamic changes of key concepts of the hungarian socialist era with natural language processing methods. *Historical Methods: A Journal of Quantitative and Interdisciplinary History*, pages 1–13.

Martina Katalin Szabó. 2015. Egy magyar nyelvű szentimentlexikon létrehozásának tapasztalatai és dilemmái. In *Segédkönyvek a nyelvészet tanulmányozásához 177*, pages 278–285. Tinta, Budapest.

György Szarvas, Veronika Vincze, Richárd Farkas, György Móra, and Iryna Gurevych. 2012. Cross-Genre and Cross-Domain Detection of Semantic Uncertainty. *Computational Linguistics*, 38:335–367, June.

James Thorne and Andreas Vlachos. 2018. Automated fact checking: Task formulations, methods and future directions. In *Proceedings of the 27th International Conference on Computational Linguistics*, pages 3346–3359, Santa Fe, New Mexico, USA, August. Association for Computational Linguistics.

James Thorne, Andreas Vlachos, Christos Christodoulopoulos, and Arpit Mittal. 2018. Fever: a large-scale dataset for fact extraction and verification. *arXiv preprint arXiv:1803.05355*.

Veronika Vincze, Martina Katalin Szabó, and Orsolya Ring. 2019. Automatic analysis of linguistic features in Communist propaganda texts. https://propaganda.qcri.org/bias-misinformation-workshop-socinfo19/paper4_final_Vincze_et_al_propaganda.pdf.

Veronika Vincze. 2013. Weasels, hedges and peacocks: Discourse-level uncertainty in wikipedia articles. In *Proceedings of the Sixth International Joint Conference on Natural Language Processing*, pages 383–391.

Veronika Vincze. 2014a. Uncertainty detection in Hungarian texts. In *Proceedings of Coling 2014*.

Veronika Vincze. 2014b. *Uncertainty Detection in Natural Language Texts*. Ph.D. thesis, University of Szeged, Szeged, Hungary.

Andreas Vlachos and Sebastian Riedel. 2014. Fact checking: Task definition and dataset construction. In *Proceedings of the ACL 2014 Workshop on Language Technologies and Computational Social Science*, pages 18–22, Baltimore, MD, USA, June. Association for Computational Linguistics.

William Yang Wang. 2017. "Liar, Liar Pants on Fire": A New Benchmark Dataset for Fake News Detection. In *Proceedings of the 55th Annual Meeting of the Association for Computational Linguistics (Volume 2: Short Papers)*, pages 422–426, Vancouver, Canada, July. Association for Computational Linguistics.

Lina Zhou, Judee K Burgoon, Jay F Nunamaker, and Doug Twitchell. 2004. Automating linguistics-based cues for detecting deception in text-based asynchronous computer-mediated communications. *Group decision and negotiation*, 13(1):81–106.

János Zsibrita, Veronika Vincze, and Richárd Farkas. 2013. magyarlanc: A toolkit for morphological and dependency parsing of Hungarian. In *Proceedings of RANLP*, pages 763–771.

Fake or Real? A Study of Arabic Satirical Fake News

Hadeel Saadany and Emad Mohamed
Research Group in Computational Linguistics
University of Wolverhampton, UK
`h.a.saadany@wlv.ac.uk`
`E.Mohamed2@wlv.ac.uk`

Constantin Orăsan
Centre for Translation Studies
University of Surrey, UK
`C.Orasan@surrey.ac.uk`

Abstract

One very common type of fake news is satire which comes in a form of a news website or an online platform that parodies reputable real news agencies to create a sarcastic version of reality. This type of fake news is often disseminated by individuals on their online platforms as it has a much stronger effect in delivering criticism than through a straightforward message. However, when the satirical text is disseminated via social media without mention of its source, it can be mistaken for real news. This study conducts several exploratory analyses to identify the linguistic properties of Arabic fake news with satirical content. It shows that although it parodies real news, Arabic satirical news has distinguishing features on the lexico-grammatical level. We exploit these features to build a number of machine learning models capable of identifying satirical fake news with an accuracy of up to 98.6%. The study introduces a new dataset (3185 articles) scraped from two Arabic satirical news websites ('Al-Hudood' and 'Al-Ahram Al-Mexici') which consists of fake news. The real news dataset consists of 3710 articles collected from three official news sites: the 'BBC-Arabic', the 'CNN-Arabic' and 'Al-Jazeera news'. Both datasets are concerned with political issues related to the Middle East.

1 Introduction

In September 2015, the Egyptian army Apache helicopters blasted a convoy of Mexican tourists in the Sinai peninsula which resulted in the killing of 12 tourists as it mistakenly took their four-wheel vehicles for terrorist targets[1]. Days later, two renowned Egyptian talk-show hosts announced that the Mexican president stated that he completely understands the Egyptian authorities' intention behind the killing of the Mexican tourists as Egypt has the right to defend its land against possible terrorist threats. A hashtag "#thankyou_Henrico_Iglassios" went viral on twitter by Egyptian pro-government twitter accounts in an attempt to show gratitude to the Mexican president for his understanding. The absurdity of the Mexican president's statement was realised a little too late. It was soon discovered that the Mexican president is not actually called 'Henrico Iglassios' and that this statement was never issued by any of the Mexican authorities who were in reality extremely angry at the incident. The source of all this commotion was an article published on an online news website, 'Al-Ahram Al-Mexici'[2]. The latter is a satirical fake newspaper that constantly posts a mockery of current events in a style that meticulously imitates the style and content of Egyptian government news agencies.

The Egyptian-Mexican incident is not the first in Egypt or in other parts of the world. There are similar incidents where satirical news is mistakenly propagated by officials or social media users because they either missed the satirical gesture or they did not investigate the source. For instance, *The Daily Beast*[3] compiled a list of nine times the headlines of the satirical online newspaper, *The Onion*[4],

---

[1]`https://www.bbc.co.uk/news/world-middle-east-34241680` (last accessed on 18 Sept 2020)
[2]`https://rassd.com/157396.htm` (last accessed on 18 Sept 2020)
[3]`https://www.thedailybeast.com/`
[4]`https://www.theonion.com/`

Proceedings of the 3rd International Workshop on Rumours and Deception in Social Media (RDSM), pages 70–80
Barcelona, Spain (Online), December 12, 2020.

fooled famous news outlets into believing their satirical stories. Among those duped were the New York Times, the US Capitol Police, and the Washington Post (Fallon, 2017). Although satirical news is essentially entertaining, it is potentially deceptive and harmful since the satirical cues could be easily missed specially if the news source is obliterated in a shared social media post or tweet. Moreover, satirical news aims to entertain by crafting unusual or surprising report of current events. According to a NY Times study on the psychology of content sharing, the motivation of almost 50% of online news sharers is to bring "entertaining content to others" (Brett, 2012). This makes the deceptive content of a satirical article more likely to be disseminated online especially when the language is a perfect parody of real news and when the original source is not mentioned.

In this research, we employ inferential statistics to investigate the defining features of Arabic satirical news on the lexico-grammatical level. We also put forward the hypothesis that Arabic satirical fake news can be automatically identified with very high accuracy. We demonstrate the effectiveness of using pre-trained word embeddings as input to statistical and neural models in detecting Arabic fake news articles with a satirical intent. To achieve its objectives, the study presents in Section 2 a brief literature review of satirical fake news detection in general and Arabic satire detection in particular. Section 3 explains the steps followed for the compilation of the corpora used in this study and the preprocessing applied to the datasets. Section 4 presents a linguistic-statistical analysis of features specific to Arabic satirical fake news. Section 5 describes the classification experiments and their evaluation. Section 6 presents an error analysis and a discussion of the classification features. Section 7 presents a conclusion on the conducted experiments and the linguistic analyses carried out in the study.

2 Literature Review

2.1 Satirical Fake News Detection

'Fake news' is generally defined as any piece of news that is intentionally crafted to present information contradicting the truth to achieve a particular goal (Gelfert, 2018). Research carried out in Natural Language Processing (NLP) and fake news in general differentiates between two types of fake news: a) satire which mimics real news, but still gives cues that it should not be taken seriously, b) intentionally crafted fake information which attempts to convince the reader of its validity such as hoax and propaganda (Rashkin et al., 2017). A major difference between satirical fake news detection and other types of fake news is that the research objective is not geared towards truthfulness estimation since the aim is fundamentally to criticise rather than to deceive. Moreover, tracking of meta-data such as news source and analyses of social network activities may not be as crucial in detecting satirical news as is the case with other types of fake news. Research on satirical fake news detection is mainly focused on writing styles, psycho-linguistic features, and atypical lexical choices which trigger a satirical cue.

An early attempt to automatically detect satirical fake news has been conducted by Burfoot and Baldwin (2009). Their contribution lies in constructing a comprehensive set of linguistic features which they present as satire-defining. Although they achieve a high classification accuracy using the standard bag-of-words approach (94.3%), they improve their accuracy by integrating three major feature sets that they discovered to be peculiar of satirical fake news. These feature sets are: 1) profanity, 2) slang, and 3) semantic invalidity. The semantic invalidity of an article is measured by how likely a named entity is mentioned in an unusual context, which is a common feature of satire (Burfoot and Baldwin, 2009).

In a more recent study, Rubin et al. (2016) proposes a SVM-based algorithm enriched with satire predicative features to classify satirical and legitimate news articles. They have shown that absurdity in an English satirical news is achieved by a number of linguistic features on the word-level such as use of slang, co-occurrence of named-entities that are not usually related and the presence of unfamiliar named-entities at the end of news articles as an ironic non sequitur (Rubin et al., 2016). The largest data for satirical fake news detection is developed by Yang et al. (2017) who utilise more than 16,000 satirical news articles collected from 14 satirical websites such as *The Onion* and *The Beaverton*[5]. The true news dataset consists of more than 160,000 articles from Google News. They develop a neural network model that analyses a hierarchy of character-word-paragraph-document. To decide which paragraph has

---

[5]https://www.thebeaverton.com/

a satirical cue, they calculate the satirical degree of a paragraph by comparing it against a learnable satire-aware vector which is based on a psycho-linguistic and stylistic feature set. Their model suggests that English satirical fake news can be predicted automatically with fairly high precision and an accuracy of up to 98.54% (Yang et al., 2017).

A further development on the dataset introduced by Yang et al. (2017) is provided by De Sarkar et al. (2018). They put forward the claim that pretrained word embeddings alone as input are sufficient in detecting satirical articles and that word-level syntax information only marginally improves classifiers performance. They point out that on the lexical level, the English satirical news article is distinct because of its use of profanity, slang, negative affect, atypical grammar structures and punctuation. They acknowledge that hand-crafted analysis of these features might contribute to a robust classifier, but a neural network model such as a Convolutional Neural Network (CNN) or a Recurrent Neural Network (RNN) trained on word embeddings is sufficient for detecting satirical fake news. Their analysis of the learned models reveals that news satire is decided by a key sentence which usually comes last. Their best classification accuracy is 98.18% achieved by an LSTM model using English Glove embeddings (De Sarkar et al., 2018).

Rashkin et al. (2017) also uses a large corpus of 14,170 satirical articles from one source, *The Onion* and add a small number of satirical articles (800) from two other sources. Their fake dataset also includes hoax and propaganda articles collected from different sources. Their model is based on hand-crafted analyses of linguistic features on the word-level. They deploy the psycholinguistic lexicon provided by the Linguistic Inquiry and Word Count software (LIWC) to categorise words into groups according to their social function (Pennebaker et al., 2001). They noticed that one distinctive feature of English satirical news is its prominent use of adverbs whereas true news uses assertive words and is less likely to use hedging words. Their model uses a Max-Entropy classifier achieving an F1 score of 0.65. They achieve a relatively lower accuracy by comparison to other satirical news detection research because their fake news dataset includes both hoax and propaganda (Rashkin et al., 2017).

2.2 Arabic Satire and Fake News Detection

Despite the fact that fake news is a major problem in the Arab world, a survey of the literature of fake news detection in NLP shows a relatively low number of research dedicated to addressing the problem in the Arabic language as compared to English. Sabbeh and Batwaah (2018) have made an attempt to automatically detect fake Arabic news by proposing a hybrid model based on content and context related features. Their model classifies a small dataset of 800 Arabic political news tweets manually labelled and collected from Twitter. A number of non-linguistic features such as account demographics, account name, number of followers are included as context-related features. They experimented with different machine learning algorithms and their best predictive accuracy was 88.9% achieved by a J48 Decision tree classifier.

A similar study for automatically measuring the credibility of Arabic web news content and specifically twitter posts is carried out by Al-Eidan et al. (2010). They present a stand-alone application for classifying Arabic tweets according to a three-level credibility measure: credible, not credible and questionable. Their system relies on document level features as well as other relevant meta-linguistic information. The main features used for analysis are: 1) similarity measures between the target document and a verified content, 2) presence or absence of inappropriate words, and 3) presence or absence of *URL* links to credible news sources. Their highest precision and recall for detecting fake news in tweets are 0.76 and 0.79 respectively. They note that their system has the best performance when using the similarity measure as the main feature for classification. (Al-Eidan et al., 2010).

Although, Arabic fake news detection research is sparse, there is a relatively large number of satire detection in Arabic NLP research. Arabic satire is studied as one form of irony where the producer is using implicit sentiment to convey a message of criticism. Karoui et al. (2017) present a supervised learning method to capture irony, satire and sarcasm in Arabic tweets. They analyse a small corpus of 5,479 of ironic and non-ironic tweets. They conclude that satire is achieved on the textual level via a number of features that create a surprising effect by breaking one norm or another. Opposite sentiments, indirect reflexive pronouns, idiomatic expressions, false assertions and exaggerated lexicon are some of the lexical features they employ to classify ironic and satirical tweets (Karoui et al., 2017). Their classification model achieves an accuracy of 72.76% by using all the features at the training stage.

Another study that examines the distinguishing characteristics of Arabic satire is conducted by Al-Ghadhban et al. (2017). They study satire as a form of sarcasm in Arabic tweets where the user conveys the opposite of what he/she means. The basic assumption behind the study is that frequency of words would be enough in defining satirical and non-satirical tweets. They train their model on a relatively small number of satirical tweets (around 340) and their model achieves an accuracy of 67.6% (Al-Ghadhban et al., 2017). As for irony detection, neural architectures have been used extensively in Arabic ironic tweet detection tasks (Allaith et al., 2019; Ranasinghe et al., 2019). A recent study by Ghanem et al. (2020) proposes a multilingual irony detection system which is able to detect irony in Arabic tweets as well as in other languages. They use both feature-based models and neural models with monolingual word embeddings. Their neural architectures (CNNs) achieved an accuracy of 80.5% for detecting Arabic ironic tweets (Ghanem et al., 2020).

It is important to mention that research in Arabic satire and irony detection mainly focuses on satirical tweets as one form of sarcasm where the author ironically sends a message opposite to his/her intention. Moreover, the analysed corpora are usually relatively small due to the difficulty of extracting tweets that are verifiably satirical. On the other hand, satirical fake news is different in two respects. First, although its objective is to entertain and ridicule, it scrupulously imitates real news register which makes it potentially deceptive. Second, it comes in the form of a news article rather than a short tweet or an online comment. To the best of the authors' knowledge, there is not a study addressing the detection of Arabic satirical news as a potential form of deceptive information. The reason might be the sparsity of the data since there are only very few Arabic news websites that are satirical in nature. The dataset analysed in the present study contributes to the creation of a new corpus for Arabic satire analysis.

3   Data Compiling and Preprocessing

The fake news dataset[6] used in the present study is a collection of 3185 crawled articles from two different sites which explicitly declare that they publish satirical fake news.[7] The first site is 'Al-Hudood'[8] (The Limits) which is a satirical online newspaper launched in 2013 by a 34-year-old Jordanian-Palestinian. It is characterised by deadpan headlines reminiscent of the US publication *The Onion*. The site averages 1 million unique visitors a month ("Al-Hudood", 2019). The second source, 'Al-Ahram Al-Mexici'[9] (The Mexican Ahram) website which is a satirical parody of the main national Egyptian newspaper 'Al-Ahram' (AlAhram-AlMexici, 2020). The two satirical online newspapers mock political leaders and Middle Eastern political events.

The real news dataset comprises 3710 articles from the BBC-News Arabic and CNN-Arabic news datasets which are available on the Sourceforge part of the Arabic Computational Linguistics project [10] as well as the Al-Jazeera News open-source dataset used in El Kourdi et al. (2004). To ensure that the fake and real datasets talk about similar themes, we chose the articles that belong to the politics section in the three real news corpora. A number of preprocessing operations were conducted on the datasets.

By surveying the corpora, it was observed that there are phrases typically repeated at the header and footer of each article and in some instances at the middle. For example, "BBC Arabic", "الشرق الأوسط" (Middle East), "شارك برأيك" (Give your opinion) are typical of the BBC News articles. The fake news dataset header and footer repeatedly mentioned phrases such as "الحدود" (Limits), "المذكور أعلاه" (Above Mentioned), "خاص للحدود" (Special for The Limits), and so on. In order to capture these stylistic guidelines, each dataset was merged into one file and frequency dictionaries of unigrams, bigrams and

---

[6]The dataset is available at `https://github.com/sadanyh/Satirical-Fake-News-Dataset`
[7]Beautiful Soup, an open-source Python Library, was used in compiling the fake dataset.
[8]`https://alhudood.net`
[9]`http://alahraam.com/`
[10]`https://sourceforge.net/projects/ar-text-mining/files/Arabic-Corpora/`

trigrams were created for each file. The top most 10% of the frequency dictionaries of each dataset included most of the website-specific phrases. Two lists of website-specific phrases indicative of the document source were manually created for each dataset and items in the lists were replaced by a space in the corpus. The cleaning of the datasets also included deletion of formatting quirks specific of each source, non-informative textual features such as special characters, English characters and diacritics. Moreover, Arabic is a quite morphologically rich language where a stem word can be packed with affixes that realise different grammatical functions. Hence segmentation of words into their constituent tokens is a very common practice in Arabic NLP research. Accordingly, the ArabicSOS, an Arabic language segmenter was employed to segment both datasets (Mohamed and Sayyed, 2019). The classification task was performed on segmented and non-segmented versions of the two datasets.

4 Linguistic-Statistical Analysis

After cleaning the data, a preliminary review of the corpus was conducted to detect category-defining features of the fake dataset. Then inferential statistic tests were performed to decide the degree of significance each of these features may have on the classification task.

4.1 Journalistic Register Measure

According to M.A.K Halliday's Systemic Functional theory of language, **register** relates to various assumptions about language in a situation type which implicates the realisation of certain selections on the lexico-grammatical stratum (Halliday, 1988). For journalistic register, the expected set of lexico-grammatical features are mainly phrases reflecting the level of news veracity such as the speaker's status, the type of news source, the venue and time of the news and so on. In Arabic journalism, there are typical phrases which are commonly used in spoken and written news publications (e.g. "من جانبه أكد", "تقرير أعده" (Reported by...), "قال الناطق باسم" (Stated the spokesman of...), (From his side, he asserted that...). Hence, a native speaker of Arabic would intuitively realise from the first few lines of a text that it is a news article if phrases typical of the journalistic register are present.

By surveying the fake news dataset, we observed that a satirical news article includes a relatively large number of journalistic cliches. A statistical analysis was, therefore, conducted to explore whether the proliferation of journalistic cliches is a defining characteristic of satirical fake news. First, a hand-crafted list of journalistic cliches mentioned in sample articles of the fake dataset was created, denoted by C. Then, the multi-set of tokens of a news article is represented by Ω. The journalistic register of a news article was measured as follows[11]:

$$J = \frac{1}{|\Omega|} \sum_{\omega \in \Omega} I_C(\omega) \tag{1}$$

where $|\Omega|$ denotes the cardinality of the multiset Ω, and I_C is the indicator function over the set C defined by:

$$I_C(\omega) = \begin{cases} 1 & \omega \in C \\ 0 & \omega \notin C \end{cases} \tag{2}$$

The function $I_C(\omega)$ takes value 1 if a cliche ω is present in an article and 0 if it is not. The journalistic register J is the aggregate number of occurrences of the journalistic cliches normalised to the article size. A two-tailed T-test with a p-value of .05 was conducted on the average means of journalistic cliches in articles of both datasets. The results were a test statistic of around 0.0628 and a p-value of 0.949, which suggests that there is no statistically significant difference between the use of these phrases in the two news categories (see Figure 1 for a probability density function plot of the journalistic register measure in both the fake and real datasets). This points to the fact that Arabic satirical news meticulously parodies real journalism register, making it difficult to identify just by using a simple typical news keyword-based approach.

---

[11] Adopted from (Burfoot and Baldwin, 2009) and adapted to our task

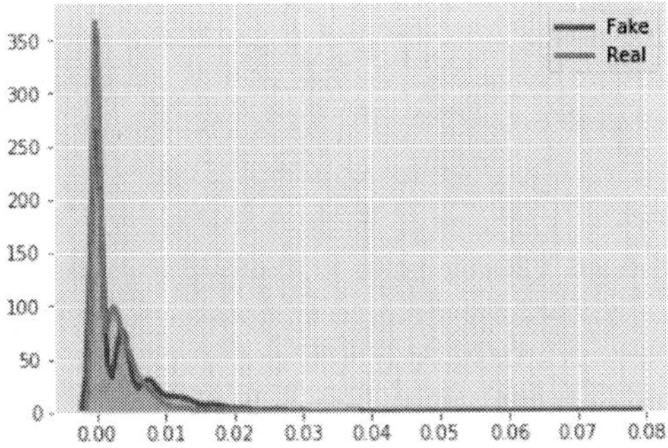

Figure 1: Probability Density Function of Journalistic Register Measure

4.2 Sentiment Intensity Measure

Professional journalists tend to express opinion objectively to avoid implied bias. By contrast, studies observed that English satirical news includes highly sentimental language for entertainment purpose (Golbeck et al., 2018; Yang et al., 2017; Rubin et al., 2016). Flouting the journalistic norm by intense affect phrases may well be considered a violation of a conversational maxim which creates satire as observed by philosophers such as Grice (Grice, 1991). Similarly, it was observed that satire in the fake news dataset is often created by an "emotional imbalance" where high positive and high negative words are present in a context of neutral journalism. To capture the above observation, the degree of sentiment intensity was measured per article against an Arabic emotion lexicon of highly positive and highly negative phrases. We used open-source emotional lexicons for Arabic language (Salameh et al., 2015; Mohammad et al., 2016a; Mohammad et al., 2016b). In an attempt to make the emotional lexicon as exhaustive as possible, a list of hand-picked highly positive and highly negative lexical items, including slang, were also added to the lexicon. The created emotion lexicon is denoted by $\mathcal{L}$. Similar to the Journalistic Register measure, sentiment score was calculated for each article as follows:

$$S = \frac{1}{|\Omega|} \sum_{\omega \in \Omega} I_{\mathcal{L}}(\omega) \tag{3}$$

where $I_{\mathcal{L}}$ is the indicator function over the set $\mathcal{L}$ defined similar to (2).

Sentiment intensity measures proved to be statistically significant (statistic ≈ 3.27, p-value ≈ 0.001) by a T-test with nan-policy='omit' option to discard all nan values in the measured arrays (the algorithm turned a nan value for words in the sentiment lexicon that did not appear in the dataset). If viewed against the journalistic register measure, it can be concluded that it is with intense sentimentality that satirical and real news diverge whereas with journalistic register they scrupulously converge.

4.3 Subjectivity Measure

Observation of the data has also suggested that fake news articles tend to be written in a more subjective tone. In Arabic, the writer can stylistically identify with the reader by using first person plural (e.g "نري" ("we think")) rather than the first person singular (e.g "أري" ("I think")). On the morphological level, first person plural is realised by the prefix "ن" (na) or the suffix "نا" (naa) which is attached to the verb. Generally speaking, news reporters refrain from using first personal pronouns to create an objective tone, unless it is included in quoted speech. In Arabic satirical news, on the other hand, it was noticed that often words such as "نروي" ("we mention"), "شارفنا" ("we are about to"), and "نأسف" ("we regret") are frequently used to report news. The satirist subjectively unites with the reader against the entity that is being targeted for criticism.

In order to explore whether this type of subjectivity is a distinctive feature of Arabic satirical fake news, the Stanford Log-linear Part-Of-Speech Tagger was used to assign parts of speech tags to tokens of the two datasets (Toutanova and Manning, 2000). The number of tokens that has a POS tag verb and starts or ends with the first person plural inflection was calculated for each dataset and normalised by the number of tokens with 'VERB' POS-tag for each dataset. The ratio of first person plural verbs in a satirical fake article to a real news article was 2.2 to 1. Thus, in a satirical news article, it is expected to find double the number of first person plural verbs than in a real one.

From the linguistic-statistical analysis, it was concluded that Arabic satirical fake news is distinguished on the lexico-grammatical level. Accordingly, representations of words as features were expected to detect the positive dataset in the classification task. In light of this, in the next section we explore different word vectorisation methods for training both statistical and neural classification models to automatically detect Arabic satirical fake news.

5 Classification Models

We used a supervised classification approach for the fake news detection task. We developed three types of models: baseline models which rely on statistical machine learning models (Section 5.1), classic machine learning models that achieved promising performance in other text classification tasks (Section 5.2) and neural models that achieved state-of-the-art results in satirical news detection (Section 5.3). We evaluated the models' performance on a held-out test set by measuring the macro average of the Precision, Recall and F1 score of the two categories (*Real* and *Fake*) as well as the total percent accuracy of each model.

5.1 Baseline Models

We used a Naive Bayes Multinomial NB classifier as a baseline model with both Bag-Of-Words (BOW) approach and TF-IDF values as numerical features. The CountVectorizer and Tfidf-Vectorizer classes from the Scikit-learn library were used in the experiment (Pedregosa et al., 2011). The max_features parameter was set to 1500 to use the most frequently occurring words as features and the max_df feature value was set to 0.7 to include words that occur in a maximum of 70% of all the documents. The classifiers were run on tokenised and segmented versions of the fake and real datasets. Additionally, the classifier with BOW values was run on n-gram vectors of a range (2,3) words and character n-grams also with range (2,3) characters. After running the experiments, it was observed that the model with Count Vectors values trained on word-level performed best with an accuracy of 96.23%. The full evaluation results are presented in Table 1.

5.2 XGBoost with TF-IDF Values

XGBoost is a version of gradient boosted decision tree classifiers. It was the winner algorithm in several classification competitions such as the KDD cup 2016 competition for predicting research relevance (Sandulescu and Chiru, 2016; Torlay et al., 2017). It has also proved to achieve high performance with an execution speed relatively faster than other classification models (Nielsen, 2016). We experimented with the XGBoost classifier using the best baseline setting. Accordingly, the XGBoost scikit-learn API was trained on the Count Vector values of article tokens in the fake and real datasets. We used the gbtree booster with a tree maximum depth of 3 to avoid over fitting. The boosting learning rate was set to 0.1 with a binary logistic objective function. With these parameters, the XGBoost classifier surpassed the best performing baseline accuracy by around 0.6%, achieving an accuracy of 96.81% (see Table 1).

5.3 Convolutional Neural Networks (CNNs)

The last experiment was conducted using a Convolutional Neural Network Architecture (CNN) which is a deep learning classification model. A CNN was chosen for two reasons: first, it has obtained state-of-the-art results on many text classification tasks in general and satirical fake news detection in particular (Zhang et al., 2015; Kim, 2014; Collobert et al., 2011; De Sarkar et al., 2018). Second, CNNs are effective in extracting morphological information and lexical patterns, both of which have been observed as characteristic features of the positive class in the current study. The input layer for the CNN was

Model	Test Set			
	Acc	P	R	F
Multinomial NB (Count Vectors)	96.23	96.47	96.23	96.24
Multinomial NB (Word-Level TF-IDF)	94.63	94.71	94.64	94.64
Multinomial NB (N-Grams)	86.08	88.03	86.09	86.22
Multinomial NB (CharLevel)	91.73	92.00	91.74	91.75
Multinomial NB (Count Vectors segmented)	95.79	95.97	95.80	95.81
Multinomial NB (Word-Level TF-IDF segmented)	94.49	94.65	94.49	94.51
XGBoost with Count Vectors	96.81	96.81	96.81	96.81
CNN with pre-trained word embeddings	**98.59**	**98.49**	**98.61**	**98.49**

Table 1: Summary of Accuracy (Acc), Precision (P), Recall (R) and F1 score (F) of classification models

pretrained fastText word-embeddings trained on Arabic Wikipedia (Bojanowski et al., 2017). The pre-trained word-embeddings covered around 92% of the corpus vocabulary. The Keras Library sequential models were employed for the task (Chollet and others, 2018). A Keras Conv1D layer was added between the Embedding layer of the pre-trained vectors and a GlobalMaxPool1D layer. The filter size for the Conv1D layer was set to 126 filters and a kernel size of 5. Training was done with Adam optimizer, ReLU and sigmoid activation functions, and a binary cross-entropy loss function. The model was initialised by 300-dimensional word-embeddings from the pretrained model and run on 10 epochs of the training data with a batch size of 10 samples for each epoch. CNN outperformed all models, resulting in an accuracy of 98.59% on the held-out test set. A summary of the evaluation report of the models is given in Table 1.

5.4 Notes on Classification Models

The results of the classification experiments show that, similar to most research studies of English satirical fake news, the Arabic satirical news can be detected with high accuracy even by the baseline models. It is worth noting that although the neural model (CNN) achieved the highest accuracy, other classifiers such as the XGBoost classifier recorded a very close accuracy figure, but with a much faster execution speed. Moreover, the precision of all the classification models was relatively higher than the recall as the number of false positive instances (i.e. a real article predicted as fake) was generally low. The CNN model with pre-trained word embeddings, however, was able to detect the fake articles that were predicted as real by other models; recording the best recall (98.61) and precision (98.49).

It can be claimed that due to its stylistic properties, lexical features per se are sufficient for accurate classification of Arabic satirical fake news. This may not be the case with other types of fake news such as propaganda and hoax where meta-linguistic data can be crucial for falsehood detection. Accordingly, even when a satirical news is disseminated in the form of an online post or a tweet without mention of its source, AI solutions can help in pointing out its deceptive content.

6 Error Evaluation and Feature Analysis

This section presents an error analysis of the false negative instances where satirical news was mistaken for real. They constituted the higher number of classification errors with all the models. We also explore the most informative features for the classification task. Both analyses provided useful insights into the problem of detecting Arabic satirical fake news.

6.1 Document-level Error Analysis

In order to have a better understanding of the classification results, we analysed some of the fake articles that were labelled "real". We noticed that the most common lexical feature of the false negative instances are the proliferation of journalistic cliches (up to 19 phrases per article), which create an almost perfect parody of a real news article. The second observation is that the misclassified articles are all written in Modern Standard Arabic (MSA) and do not contain any slang items. However, there were MSA phrases carrying exaggerated sentiment import that are not common in journalistic register. For example, in one misclassified article, the leader of the Hezbullah Shia political party is described as the "all-time comrade" (رفيق دربه) and the "resistance armour" (سلاحه في المقاومه) of the leader of the Houthi Movement (a Shia Yemeni Movement involved in the current Yemeni war). Most of the high sentiment lexical items in the misclassified articles were not captured by the emotion lexicon used to measure sentiment intensity. They were mostly coined hyperboles realised on the lexico-grammatical level by common nouns or descriptive adjectives that were crafted by the satirist.

6.2 Word-level Error Analysis

An attempt was made to find the most effective vectors for the baseline classification models, the CountVectorizer and TF-IDF. This was achieved by zipping the baseline classifier coefficients with the feature names. Accordingly, the log of the estimated probability of a feature given the positive class (fake news) was extracted for the top 30 'most fake' and top 30 'most real' words in the two datasets.

A number of observations were made on the basis of the analysis of the most-informative feature list. First, the most prominent feature name for fake news is "المصدر" (the source). The veracity of the news source is overemphasised in fake news in order to highlight the reliability of the news story and maximise the satirical effect. Similarly, journalistic cliches such as "asserted" and "the spokesman" always come up as features of the fake class. Again, the objective is to envelop satire with an aura of credibility. Along with credibility markers, high sentiment polarity words are among the most prominent features (e.g. "عفيفة" (chaste), "أحمق" (foolish), "أجود" (most generous)). Moreover, names of a number of political figures such as Trump and Al-sisi (president of Egypt) are at the top of the fake features. This is quite logical since divisive political figures ignite a rich material for satire.

As for the real news dataset, the most informative features are mainly phrases denoting time and venue of the news (e.g. "الاثنين" (Monday), "الماضي" (last), "غزة" (Gaza)). Documenting the news report is of higher importance in real news articles. Also, contrary to fake news, with different classification runs, the most informative features of the real news dataset did not once contain an emotive phrase.

7 Conclusion

This paper proposed a novel classification problem for Arabic computational linguistics and machine learning: detecting whether an Arabic news article is true or satirical. The study experimented with different classification techniques including state-of-the-art deep learning models. Experimental results showed a superior performance of a proposed architecture employing the combination of a CNN with pretrained word embeddings. A linguistic-statistical analysis of the satirical dataset gave a deeper insight into the phenomenon of Arabic satirical news. However, the high accuracy of the CNN neural architecture suggests that pre-trained word embeddings as input, without the aid of any syntactic information or any hand-crafted linguistic features, suffices in capturing Arabic satire. The reason is that a satirical cue is always there in a fake satirical article. Although the writer's aim is for the absurdity of the news to be ultimately realised, the satirical message can still be missed and taken for real. It is worth mentioning, that the classifier successfully detected the deceptive content of the satirical fake news article published by Al-ahram Al-Mexici which was the cause of the diplomatic crisis between Egypt and Mexico mentioned in the introduction of this paper. The study gives promising evidence that the dissemination of this type of deceptive content via the internet can indeed be automatically prevented by AI solutions.

References

Rasha M BinSultan Al-Eidan, Hend S Al-Khalifa, and AbdulMalik S Al-Salman. 2010. Measuring the credibility of Arabic text content in Twitter. In *2010 Fifth International Conference on Digital Information Management (ICDIM)*, pages 285–291. IEEE.

Dana Al-Ghadhban, Eman Alnkhilan, Lamma Tatwany, and Muna Alrazgan. 2017. Arabic sarcasm detection in twitter. In *2017 International Conference on Engineering & MIS (ICEMIS)*, pages 1–7. IEEE.

"Al-Hudood". 2019. The limits. https://alhudood.net. [Online; accessed 29-June-2019].

AlAhram-AlMexici. 2020. The mexican ahram. http://alahraam.com/. [Online; accessed 29-June-2020].

Ali Allaith, Muhammad Shahbaz, and Mohammed Alkoli. 2019. Neural Network Approach for Irony Detection from Arabic Text on Social Media. In *FIRE (Working Notes)*, pages 445–450.

Piotr Bojanowski, Edouard Grave, Armand Joulin, and Tomas Mikolov. 2017. Enriching word vectors with subword information. *Transactions of the Association for Computational Linguistics*, 5:135–146.

Brian Brett. 2012. The psychology of sharing: why do people share online. *New York Times*.

Clint Burfoot and Timothy Baldwin. 2009. Automatic satire detection: Are you having a laugh? In *Proceedings of the ACL-IJCNLP 2009 conference short papers*, pages 161–164.

François Chollet et al. 2018. Keras: The python deep learning library. *Astrophysics Source Code Library*.

Ronan Collobert, Jason Weston, Léon Bottou, Michael Karlen, Koray Kavukcuoglu, and Pavel Kuksa. 2011. Natural language processing (almost) from scratch. *Journal of machine learning research*, 12(Aug):2493–2537.

Sohan De Sarkar, Fan Yang, and Arjun Mukherjee. 2018. Attending sentences to detect satirical fake news. In *Proceedings of the 27th International Conference on Computational Linguistics*, pages 3371–3380.

Mohamed El Kourdi, Amine Bensaid, and Tajje-eddine Rachidi. 2004. Automatic Arabic document categorization based on the Naïve Bayes algorithm. In *proceedings of the Workshop on Computational Approaches to Arabic Script-based Languages*, pages 51–58.

Kevin Fallon. (2017). Fooled by the onion: 9 most embarrassing fails. [Online; accessed 15-July-2019].

Axel Gelfert. 2018. Fake news: A definition. *Informal Logic*, 38(1):84–117.

Bilal Ghanem, Jihen Karoui, Farah Benamara, Paolo Rosso, and Véronique Moriceau. 2020. Irony detection in a multilingual context. In *European Conference on Information Retrieval*, pages 141–149. Springer.

Jennifer Golbeck, Matthew Mauriello, Brooke Auxier, Keval H Bhanushali, Christopher Bonk, Mohamed Amine Bouzaghrane, Cody Buntain, Riya Chanduka, Paul Cheakalos, Jennine B Everett, et al. 2018. Fake news vs satire: A dataset and analysis. In *Proceedings of the 10th ACM Conference on Web Science*, pages 17–21. ACM.

H Paul Grice. 1991. *Studies in the Way of Words*. Harvard University Press.

Michael AK Halliday. 1988. On the ineffability of grammatical categories. In *Linguistics in a systemic perspective*, page 27. John Benjamins.

Jihen Karoui, Farah Banamara Zitoune, and Veronique Moriceau. 2017. SOUKHRIA: Towards an Irony Detection System for Arabic in Social Media. *Procedia Computer Science*, 117:161–168.

Yoon Kim. 2014. Convolutional neural networks for sentence classification. *arXiv preprint arXiv:1408.5882*.

Emad Mohamed and Zeeshan Ali Sayyed. 2019. Segmenter and Orthography Standardazier (SOS) for Classical Arabic (CA). https://https://github.com/zeeshansayyed/ArabicSOS.

Saif M. Mohammad, Mohammad Salameh, and Svetlana Kiritchenko. 2016a. How translation alters sentiment. *Journal of Artificial Intelligence Research*, 55:95–130.

Saif M. Mohammad, Mohammad Salameh, and Svetlana Kiritchenko. 2016b. Sentiment Lexicons for Arabic Social Media. In *Proceedings of 10th edition of the the Language Resources and Evaluation Conference (LREC)*, Portorož, Slovenia.

Didrik Nielsen. 2016. Tree boosting with xgboost-why does xgboost win" every" machine learning competition? Master's thesis, NTNU.

F. Pedregosa, G. Varoquaux, A. Gramfort, V. Michel, B. Thirion, O. Grisel, M. Blondel, P. Prettenhofer, R. Weiss, V. Dubourg, J. Vanderplas, A. Passos, D. Cournapeau, M. Brucher, M. Perrot, and E. Duchesnay. 2011. Scikit-learn: Machine learning in Python. *Journal of Machine Learning Research*, 12:2825–2830.

James W Pennebaker, Martha E Francis, and Roger J Booth. 2001. Linguistic inquiry and word count: Liwc 2001. *Mahway: Lawrence Erlbaum Associates*, 71(2001):2001.

Tharindu Ranasinghe, Hadeel Saadany, Alistair Plum, Salim Mandhari, Emad Mohamed, Constantin Orasan, and Ruslan Mitkov. 2019. RGCL at IDAT: deep learning models for irony detection in Arabic language. *Working Notes of FIRE 2019 - Forum for Information Retrieval Evaluation*, pages 416–425.

Hannah Rashkin, Eunsol Choi, Jin Yea Jang, Svitlana Volkova, and Yejin Choi. 2017. Truth of varying shades: Analyzing language in fake news and political fact-checking. In *Proceedings of the 2017 Conference on Empirical Methods in Natural Language Processing*, pages 2931–2937.

Victoria Rubin, Niall Conroy, Yimin Chen, and Sarah Cornwell. 2016. Fake news or truth? using satirical cues to detect potentially misleading news. In *Proceedings of the second workshop on computational approaches to deception detection*, pages 7–17.

Sahar F Sabbeh and Sumaia Y Batwaah. 2018. Arabic news credibility on twitter: An enriched model using hybrid features. *Journal of Theoretical & Applied Information Technology*, 96(8).

Mohammad Salameh, Saif Mohammad, and Svetlana Kiritchenko. 2015. Sentiment after translation: A case-study on Arabic social media posts. In *Proceedings of the 2015 conference of the North American chapter of the association for computational linguistics: Human language technologies*, pages 767–777.

Vlad Sandulescu and Mihai Chiru. 2016. Predicting the future relevance of research institutions-the winning solution of the kdd cup 2016. *arXiv preprint arXiv:1609.02728*.

L Torlay, Marcela Perrone-Bertolotti, Elizabeth Thomas, and Monica Baciu. 2017. Machine learning–xgboost analysis of language networks to classify patients with epilepsy. *Brain informatics*, 4(3):159–169.

Kristina Toutanova and Christopher D. Manning. 2000. Stanford log-linear part-of-speech tagger 3.9.2.

Fan Yang, Arjun Mukherjee, and Eduard Dragut. 2017. Satirical news detection and analysis using attention mechanism and linguistic features. *arXiv preprint arXiv:1709.01189*.

Xiang Zhang, Junbo Zhao, and Yann LeCun. 2015. Character-level convolutional networks for text classification. In *Advances in neural information processing systems*, pages 649–657.